The Sternberg Fossil Hunters

The Sternberg Fossil Hunters

A Dinosaur Dynasty

Katherine Rogers

Mountain Press Publishing Company
Missoula, Montana
1991

Cover painting by Dorothy Sigler Norton

Library of Congress Cataloging-in-Publication Data

Rogers, Katherine, 1910-
    The Sternberg fossil hunters: a dinosaur dynasty / Katherine Rogers.
       p.   cm.
    Includes bibliographical references and index.
    ISBN 0-87842-300-1 : $10.00
    1. Sternberg, Charles H. (Charles Hazelius), b. 1850.
  2. Dinosaurs—Research—West (U.S.)—History—Popular works.
  3. Dinosaurs—Research—Alberta—History—Popular works.
  4. Paleontologists—United States—Biography—Popular works.
  I. Title.   II. Title: Dinosaur dynasty.
  QE707.S8R64   1991
  560.9—dc20
  [B]   91-25895
                                           CIP

MOUNTAIN PRESS PUBLISHING COMPANY
P.O. Box 2399 • 2016 Strand Avenue
Missoula, Montana 59806
(406) 728-1900

Table of Contents

Maps

Foreword

Who finds fossils? What sort of people let dreams of archaic worlds dominate their lives?

A special kind of explorer seeks out Nature's burial grounds, a special kind of hunter aims to bring his quarry back to life. Most of the early fossil hunters worked in anonymity, but some became models for cinema stereotypes. These miners of natural history include Carl Akley, Frank ("Bring 'em back alive") Buck, Roy Chapman Andrews and Barnum Brown. Historically, however, the Sternbergs—all four of them—made perhaps the most enduring impact. These fossil-finding fanatics earned their bread and butter by collecting and selling the tailings of long extinct, but in their minds, still vividly living, organisms.

This book centers on three generations of Sternbergs and their odyssey through the 19th and early 20th centuries and backwards millions of years into ages long past.

The account begins with Levi Sternberg, a Lutheran and then Presbyterian minister, the family patriarch. The story of Levi and his family is coincident with the westward move of settlers. Caught up in the winning of the West are two of his sons, who also figure most prominently in this epic of digging for fossils. The eldest of Levi's sons was George Miller Sternberg, who eventually became the surgeon general of the United States. Charles Hazelius Sternberg was younger and often aided by his more politically active older brother. The peripatetic Charles founded the dynasty whose exploits are recounted here. All three of his sons also had careers in paleontology. The eldest was George Fryer, named for his uncle. His younger brothers, Charlie and Levi, figured prominently in Canadian paleontology. George, on the other hand, spent most of the last few decades of his life associated with Fort Hays State University in Hays, Kansas, promoting the fossils of his native state.

Katherine Rogers, the author of this volume, knew George F. Sternberg well. Associated with the Sternberg Memorial Museum, even before it came to be known as such, she has a unique and very personal perspective of the Sternbergs. Rogers, a retired journalism professor from Fort Hays State University, has had access to all of George F. Sternberg's field diaries, letters, and records. These resources have provided her with a sense of the privations of paleontological field work during the late 1800s and early 1900s.

With these materials close at hand, it is not surprising that Rogers has generated an account filled with the kind of detail rarely revealed in similar second-hand narratives. Her approach renders this history almost novelistic. Often the reader will be impressed with how much of the Sternberg personality is in evidence. Charles H. Sternberg was a significant player in the famous Cope/Marsh feuds. Rogers is able to present a balanced view of Sternberg's impact and involvement in this distasteful scientific interlude.

Various Sternbergs prospected for fossil remains throughout Western America, Canada, and even Patagonian South America. Travel between localities was long, risky, involved, and probably tedious. Nevertheless, the rich fossil fields they explored are still yielding important finds. The Sternbergs were ground-breakers in both a literal and figurative sense. They often were the first scientifically sanctioned diggers in a new area. Some of the New World's most important fossil-bearing regions were first explored by them. Subsequent generations continue to return to their sites or to areas nearby. Material that the Sternbergs uncovered continues to be prepared and studied today, assuring their scientific immortality.

This volume is not intended to be the definitive scholarly biography of the Sternberg fossil hunters. Rogers has, instead, provided an excellent and readable account of a family united in a truly unique lifestyle—one that no longer exists. The "species" that the Sternbergs represent is as extinct as the fossilized creatures they uncovered. And yet, through their scientific contributions, they have achieved an immortality of sorts. Their legacy is fossils—fossils that enrich museums around the world, fossils that will open eyes of wonder to imagined scenes of ancient battles waged in surreal prehistoric landscapes of the imagination.

Dr. Charles R. Crumly
Department of Biology
San Diego State University

Preface

This book brings together for the first time stories of the lives of all of the Sternberg fossil hunters. Charles H. Sternberg wrote two books in the early 1900s about his activities, and paleontologists and those writing about fossil hunters continue to make frequent reference to him and to his sons; but the whole story of their adventures has never been told.

I knew Charles's oldest son, George F., for forty years while he was curator of the museum at what is now Fort Hays State University, Hays, Kansas. I was first a student, then a faculty member at the university, teaching journalism and writing features and press releases about the museum. Originally, this book was to be the story of George's life, as requested by the president of the university. But as research progressed, I realized a more comprehensive story should be written.

My first source of help was Myrl Walker—George Sternberg's friend, assistant, and successor as curator. Our numerous interviews brought forth stories and data previously unrecorded. Walker knew many members of the Sternberg family, had delved deep into their genealogy and history, and worked side by side with George Sternberg for most of the years Sternberg was in Hays. Walker hoped and planned to write this book himself, but ill health intervened. He died soon after sharing the Sternberg materials with me, and at that time I was given temporary custody of all the Sternberg files, including what remained of George's diary from two years in Patagonia. Handwitten in pencil on 3" by 5" sheets of notebook paper, tied together with string, the diary provided the only firsthand information about this chapter in Sternberg's life.

I also received all of Walker's research material from Yale University library concerning the Sternbergs, and voluminous files about all of the Sternbergs. Eight photo albums contained 2,500 photographs, most of which were taken by George between 1906 and 1965. From the

captions I could recreate incidents and scenes; but hundreds of the pictures have no captions and have been almost useless as resource material. Unless otherwise noted, the photographs that illustrate this book are taken from George Sternberg's albums.

This is primarily a story of the men and their lives, not a scientific study. I wanted to preserve the flavor and style of Charles H. Sternberg. Sometimes his directions were not explicit, his geography not always perfect, and his descriptions of fossils not always accurate. Some of the lakes and streams he named have disappeared or changed course, so are no longer geographically accurate. He spoke often of Elkader, Kansas and Steveville, Alberta, Canada, where Sternbergs established camps. The towns are not to be found on today's maps. Since some county names have changed, the maps may not be entirely accurate; they are intended only to provide a general idea of where the expeditions took place. Scientific terminology has also changed in some instances. For example: the Sternbergs found a number of enormous fish skeletons—*Portheus*, they said; now such species are known as *Xiphactinus*. I wrote this book for average readers, persons interested in true adventure accounts of incidents with Indians, the violence of nature, living in tents, and tramping thousands of miles across much of western United States, parts of Canada and southern Argentina.

The Sternbergs earned their livelihood and supported their families by finding and selling fossils of prehistoric animals, and their expertise and fossils brought international fame. From its beginning in about 1870, the Sternberg dynasty lasted for more than a century before it died out like the dinosaurs the Sternbergs hunted. Sternberg fossils are found in museums in a dozen foreign countries and at least twenty-two states.

Acknowledgements

I am indebted to friends and associates of George Sternberg, not only Myrl Walker, but William ("Bill") Eastman who also worked with George and remembered many lively incidents. He and Standlee V. Dalton, a botany professor at the university, often accompanied George on weekends and were on hand to help take out the huge *Portheus* "fish within a fish." Charles W. Sternberg, George's son and sole survivor of the "Sternberg Dynasty," contributed facts and reminiscences from his childhood years with his father.

Valuable assistance came from Canadian museums: Dr. Loris Russell and staff members of the Royal Ontario Museum, Provincial Museum of Alberta at Edmonton, the Tyrrell Museum of Palaeontology at Drumheller, and the Canada National Museum of Natural Sciences at Ottawa. Dr. William D. Turnbull, curator emeritus of the Field Museum of Natural History of Chicago, and his staff supplied official records of George Sternberg's expedition to Argentina. The National Museum of Natural History at the Smithsonian Institution, the American Museum of Natural History in New York, Jack Murphy and the staff of the Denver Museum of Natural History, Dr. Charles R. Crumly and Dr. Robert Sullivan of the San Diego Museum of Natural History—all contributed to help make this book accurate and authentic.

Help came from Hartwick College, Oneonta, New York, which made available material from Hartwick Seminary, defunct since 1945; the Oregon State Historical Society, Kansas State University Archives; Kansas State Historical Society; Frick Memorial Museum, Oakley; Historic Fort Hays, Sternberg Memorial Museum; Hays Public Library and Forsyth Library at Hays; photographer Charlie Riedel and map maker Jack Jackson; Dr. Jerry Choate, Dr. Richard Zakrzewski, and Jay Burns of Sternberg Museum; Dr. Michael Nelson, Dr. Raymond Wilson, Jerry Green, Fr. Blaine Burkey for editorial and marketing help. Thanks are due to Mike Corn for proofreading and technical help. I also acknowledge Elizabeth Noble

Shor for her book *Fossils and Flies*; and Douglas Preston for his *Dinosaurs in the Attic* and his article *Sternberg and the Dinosaur Mummy*.

Last, but certainly not least, I thank my daughter Jean Weber for her editorial expertise, constructive criticism, and computer competency; and thanks, too, to the rest of my family who listened patiently and provided encouragement through my 10 years of research and writing, and who respected "George's Room," that portion of my home which held all the research materials and was therefore out-of-bounds to the family. Without such help, the book could not have been completed.

Sternberg Genealogy

Levi Sternberg was one of eleven children, representing the fourth generation of descendents of Johan Jacob Sternberg (born 1665) and Catherine (born 1667), who came to New York on the Ship Lyons in 1710.

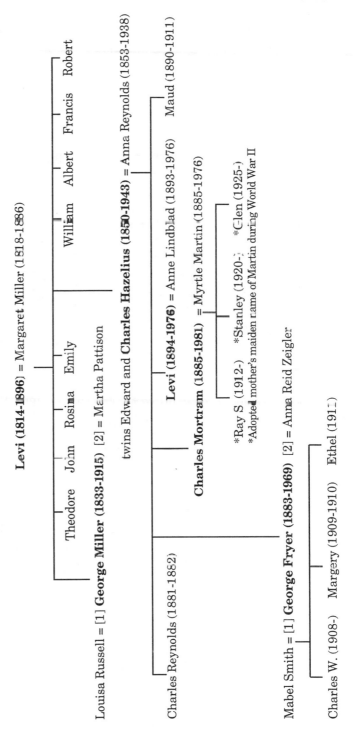

Levi (1814-1896) = Margaret Miller (1818-1886)

Louisa Russell = [1] **George Miller (1833-1915)** [2] = Martha Pattison

Theodore John Rosina Emily

twins Edward and **Charles Hazelius (1850-1943)** = Anna Reynolds (1853-1938)

William Albert Francis Robert

Levi (1894-1976) = Anne Lindblad (1893-1976)

Maud (1890-1911)

Charles Mortram (1885-1981) = Myrtle Martin (1885-1976)

*Ray S (1912-) *Stanley (1920-) *Glen (1925-)
*Adopted mother's maiden name of Martin during World War II

Charles Reynolds (1881-1882)

Mabel Smith = [1] **George Fryer (1883-1969)** [2] = Anna Reid Zeigler

Charles W. (1908-) Margery (1909-1910) Ethel (1912-)

PROLOGUE

The grimy little nine-year-old grabbed his father's chisel and hammer, dropped to his knees, and bent low over a flat chunk of chalk rock. Without giving thought to any design, he started chipping a simple inscription:

G.F.S.
1892   Age 9

The child was George Fryer Sternberg, eldest son of Charles Hazelius Sternberg, and he had just found a large fossil. It was his first discovery; his excitement exceeded all bounds, for he knew he had pleased his father. Now he wanted to record his achievement.

Charles H. Sternberg had hunted fossils since he was seventeen years old and now at age forty-two he wanted, more than anything else in life, his sons to feel the same overpowering determination he felt to uncover the secrets of prehistoric life. Charles had first exposed little George to fossil hunting when the child was six years old, and that episode had been little short of a disaster. But today he dared to believe his efforts had not been in vain. The boy had taken the initiative to look for fossils in the chalky cliffs near the family camp, and had found what proved to be an excellent specimen of a plesiosaur.

Little George laid down his father's chisel, rose from his squatting position, and smiled triumphantly at his father, then at his mother, Anna, who stood close by, holding the hand of his younger brother Charles M.

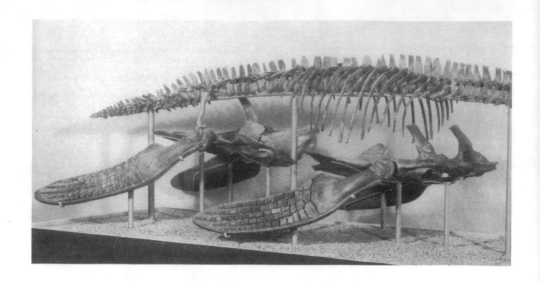

Conflicting emotions showed in Anna's face. She was proud of George, glad he had pleased his father, and happy for the joy she saw in the child's face. But as she gripped the hand of her younger son, she could not suppress the apprehension she felt. Was fossil hunting to become a family affair, an unending compulsion that would always take precedence over normal activities and routine of home life? Here she was, standing in the blistering heat of a Kansas sun in midsummer, battling the incessant wind, living in a tent pitched on the barren wasteland of southern Logan County in far western Kansas, without the normal amenities of home. Even water had to be hauled from twenty miles (32 km) away. All this she and the boys endured, just to have few weeks in summer with Charles. Normally, she and the children would be in Lawrence at the other end of the state, on the banks of the Kansas River, in a pretty little city with tree-lined streets and large, comfortable homes. Anna had lived in Lawrence since childhood, and after her marriage to Charles, continued to live with her parents since her husband was gone much of the year. He spent winter months, however, working in his "laboratory" cleaning and preparing his fossils for sale.

Fossil hunting was Charles's business and the only source of income for his family. It was a unique but legitimate occupation, which in the late nineteenth century involved only a handful of people who roamed the rocky expanses of western North America in search of fossilized remains of prehistoric plant and animal life. Even fewer persons made fossil hunting and selling their only occupation. Charles Sternberg had declared his life's intent while still a teenager, and when Anna came into his life he was already thirty years old and well-established in his profession. She dutifully cared for her children and

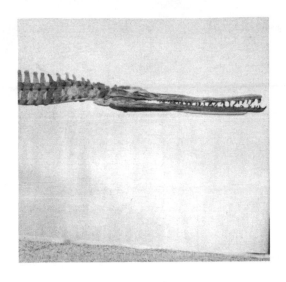

A small plesiosaur, similar to the one found by George Sternberg at age 9. This specimen is in the Sternberg Memorial Museum.

accepted the increasing burden of almost sole responsibility for their training. This summer's expedition to the chalk beds living in a tent on the hot prairie would be a once-in-a-lifetime experience, as far as she was concerned.

For Charles, George's discovery was epochal. The fossil specimen was large and in excellent condition, but more significant was the fact that the child had shown he had a natural talent for finding fossils, and in the years ahead might become a partner in his father's business. The summer had not been highly successful for Charles. He desperately needed to secure a good specimen or two to justify his months of work. So it was with relief and pride that he reverently exclaimed when he looked at the bones his son found protruding from the chalky ledge: "George, thank God, you have found a plesiosaur." Charles was a devoutly religious man and never failed to thank the Almighty for any good fortune that came his way.

The plesiosaur was crated in a sturdy box, packed with dried grass and excelsior, and hauled twenty miles (32 km) to the Kansas Pacific Railroad station at Buffalo Park for shipment to Lawrence. Eventually it was sold; just one of thousands of fossils Charles Sternberg shipped to museums all over the world in his sixty-year career.

For little George, this was one of the most important discoveries of his life and he memorialized it with a hunk of rock that he left at the discovery site. Many years later, friends carefully removed the marker and gave it a place of honor in the Sternberg Memorial Museum, less than a hundred miles (160 km) from the point of discovery.

But in the interim, the Sternbergs—the father and three sons— became a family of fossil hunters with an international reputation for their discoveries and achievements.

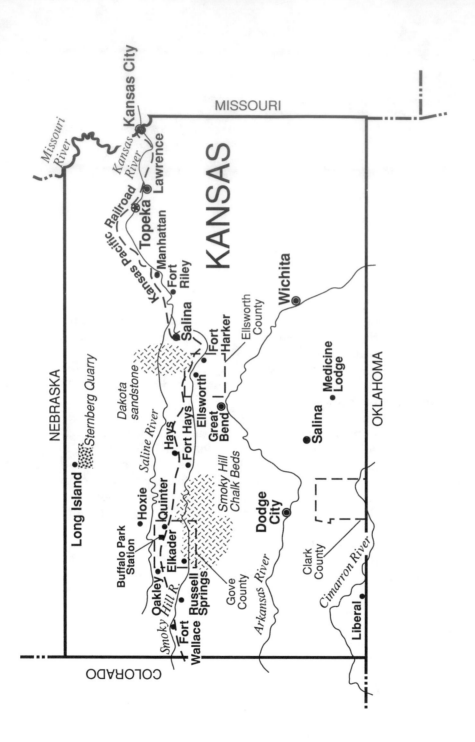

CHAPTER ONE

Manifest Destiny

C harles Sternberg shifted his gaze from the bones of the ancient reptile to his radiant son George, and his heart filled with emotion and pride. Here was a son to be proud of, a natural-born fossil hunter!

Memories flooded back of a time long ago when he himself as a small boy proudly showed his father some shells and strange stones. These discoveries, however, had evoked neither excitement nor praise from his father, only cutting remarks about how Charles was wasting time playing with old rocks and shells. How natural it was, Charles mused, for a child to want to share his finds with a parent and to expect approval in response. Charles had never experienced much approval from his father, Levi; and his mother, Margaret, spoke quietly to her son about his discoveries and did so without compromising the stern discipline of the Sternberg home.

Levi Sternberg was now an old man, and Charles respected him, even though they had never achieved a warm parent-child relationship. His childhood years seemed to Charles like the fossils he unearthed, buried deep in the rock and dust of the past.

Charles spent his childhood in New York State where the Sternbergs lived in the valley of the Susquehanna River in Otsego County, not far from Cooperstown. Levi was a deeply religious man, devoted to the education and religious training of his family. It was the duty of his wife to cope with the day-to-day trivia associated with a large family: Charles and his twin brother, Edward, were numbers six and seven of eleven children.

Levi enrolled all his sons[1] in Hartwick Seminary at age six. Levi's own education had begun there in 1820 when he was that age. Margaret's father, Dr. George Miller, was a theology professor at the

Levi Sternberg was principal of Hartwick Seminary from 1850 to 1865, and most of his eleven children received their early education here. —Hartwick College Archives, Oneonta, N.Y.

seminary; and when Levi completed college, he succeeded his father-in-law on the seminary faculty. Later, he became principal of the school.

The discipline at Hartwick was strict; the curriculum, classical; and the hours, long. Nevertheless, Charles and his siblings found time to romp through the hills and meadows and to picnic under the hickory, pine, and hemlock trees there. Charles happily searched the rock for unusual and interesting examples of nature's bounty. He cut out shells he found in the limestone strata and took them with a fistful of wildflowers to show his mother who shared his enthusiasm, if not his interest. Loving and caring, she made time to draw each child close to her and create a lasting bond of confidence.

The thought of his mother brought a wave of nostalgia to Charles now, as he glanced at his own family clustered around him in the hot, dusty sun of a Kansas summer day. He recalled the time when, at ten years old, he had an accident that affected him permanently. While chasing an older boy through the upper level of the family farm's huge barn, he fell twenty feet (6 m) through a hole at the top of a ladder and landed unconscious with an injured leg on the floor below. The older boy carried him to his mother; and the long recovery period drew him closer to her than ever. The family physician believed the injury was only a sprain and applied a bandage. Later, doctors discovered that the fibula of the left leg was injured and the left knee, dislocated. Charles was forced to use crutches, but even so he still roamed the hills and tried to keep pace with the other children. He threw away his crutches as soon as possible, but walked with a limp for the rest of his life.

Charles would have enjoyed sharing little George's discovery with his own mother, but Margaret Sternberg was no longer living. Charles recalled her support when as a small boy he had brought his discoveries to her. He could do no less for little George. Of course, it would be wonderful for George to become a real fossil hunter—and Charles vowed to do all in his power to bring this to pass.

Charles remembered the time when he and the other Sternberg children were playing in the attic of his Uncle James's house and found a cradle filled with fossil shells and quartz crystals. Fascinated, Charles fingered each piece, studying the markings, coloring, and uniqueness as he spread them out on the attic floor. Uncle James gave the cradle to the boy, explaining that his brother had found the contents some years ago and "fortunately had died early enough in life that he had brought no disgrace upon the family by wasting his time in such idle play as rock gathering."[2] The collection provided Charles with hours of entertainment and study. Eventually, the family moved and he had to part with it; but he entrusted his beloved collection to a favorite aunt "for safekeeping." As an adult, he tried to find out, but never did, what became of his childhood treasure.

On the cradle-collection items that looked as though they might have been alive at one time, Levi commented that the Almighty, who created the rocks, could easily, at the same time, have created the ancient plants and animals as fossils, just as they were found.[3]

Levi never had time nor patience for anything foolish. He was a minister, scholar, educator, and a man of firm convictions. His aggressiveness and adherence to his own opinions sometimes got him into trouble, and were, in fact, underlying reasons why the Sternbergs left New York and wound up in Kansas.

As principal of Hartwick Seminary, Sternberg made drastic changes. He admitted female students, hired a female faculty member, and inaugurated courses in vocational agriculture and normal training for teachers. These innovations did not please the administrative board, and their disapproval quickly became intolerable to Levi.

Levi Sternberg, as he appeared when principal of Hartwick Seminary, the oldest Lutheran school in America. —Hartwick College Archives, Oneonta, N.Y.

The two oldest Sternberg children—George Miller (named for his grandfather) and Theodore—had finished school at Hartwick, and George had gone on to earn a medical degree from Columbia College (now Columbia University). Shortly after opening a private practice, he enlisted in the Union army and was joined by Theodore.

Charles was fifteen at the time, old enough to realize what was happening to his family and eager to discover what lay ahead. He felt particularly close to George, although his brother was twelve years his senior, and wished many times for a chance to talk meaningfully with him. George bridged the gap between Levi and Charles in discussions about the theories of Charles Darwin or about the fossils being discovered in the excavations along the recently widened Erie Canal, not far north of Cooperstown.

As a Lutheran minister, Levi had little use for Darwin's new *Origin of the Species* (1859), and scoffed at any suggestion that departed from the fundamentalist theory of the Creation. On this issue, more than any other, young Charles and his father found a common denominator for discussion, as they sought to reconcile the differences between the scientific and the fundamentalist theory of the Creation.[4]

Proponents of Darwin's theory reactivated news stories of an incident nearly 75 years earlier when Gen. George Washington had

8

received a request from the English in 1782. They asked him to permit Dr. Christian Friedrich Michaelis to visit the farm of a Rev. Annan in the highlands of the Wallkill Valley, near the southern border of New York, in order to examine giant teeth that had been dug up there. Benjamin Franklin became intensely interested in these and other fossil teeth and bones and said it appeared to him that animals capable of carrying such large heavy tusks must themselves be large creatures, too bulky to pursue and take prey. He was inclined to think the "knobs" or teeth were only a variation from the norm. Thomas Jefferson believed God would never allow a species to become extinct. He concluded that creatures having such teeth must still be roaming the unexplored West.[5]

Charles had read about Cotton Mather, who, many years earlier, decided that the big teeth and leg bones found in far-flung places were from godless giants wiped out in the flood of Noah's time. Discoveries of such bones in various parts of the earth posed no problem for Mather and his friends, who simply explained that the Ark covered great distances in its day.[6] Charles also remembered learning that in the 1760s molars and pieces of huge tusks were found at Big Bone Lick on the Ohio River south of the Ohio-Kentucky border. Some philosophers explained these finds as fragments of skeletons of the animals who died on Noah's Ark and were buried at sea. But one man mused: "That was a right smart boat. Just forty days, but she tossed out carcasses from Gibraltar to Pennsylvania, and then all the way back to Mount Ararat. John Paul Jones could have used a craft that pert, 'stead of them rotten hulks the Frenchies give 'im."[7]

While Charles mulled over these stories and his father's problems, he wished George would come home. He knew George would know how to talk with his father and help Levi make a difficult decision. Without the confidence and support of the seminary board, Levi could no longer effectively head his beloved Hartwick. He resigned in 1865 and accepted a position in Albion, Iowa, to teach at Iowa Lutheran College. Charles didn't know much about Iowa, but he did remember reading a few stories about Indians in the area. Maybe it would be exciting to live there. Albion was a small town in the heart of the state, northeast of Des Moines. Maybe that was the "Wild West!"

With heavy hearts, Levi and Margaret packed up their belongings and said good-bye to their many relatives in New York. The Sternberg clan had lived in New York since coming to America in 1710. Iowa was a long way west; but for the nine children—ranging in age from twenty-two to five—the move would be a great adventure. The Civil War had ended, but George and Theodore had not rejoined the family—at least, not right away. Theodore had many ideas for his own future and was not anxious to return home; George decided to make army medicine his career.

After the war, George was assigned as executive officer at one army hospital after another, and about the time the rest of his family moved to Iowa, he was ordered to report to Jefferson Barracks, near St. Louis. Here, he wouldn't be far from Iowa and his family, but would be a long way from Maria Russell, his childhood sweetheart. She had waited for him in Cooperstown while he served with the Army of the Potomac and was taken prisoner of war at the Battle of Bull Run. His medical care of the wounded of both armies, escape from prison, and return to active service gave him an impressive record. He expected his assignment to Jefferson Barracks to be somewhat permanent, so began making plans to marry Maria. They were wed on October 19, 1865, and went immediately thereafter to Jefferson Barracks. Six months later Sternberg was transferred to Fort Harker, Kansas, to serve as post surgeon.

Fort Harker was one of many small outposts established to assist the ever-increasing flow of settlers moving west. Cheyenne Indians

Wedding picture of Dr. George M. Sternberg, taken in Rome, 1865.
—Ellsworth Kansas Historical Society

Fort Harker, Kansas, as painted by Herman Stieffel, Company K5 Infantry, soldier artist of the West, 1871. —from painting at Fort Harker Museum

still roamed the prairies, plundering small groups of wagons, raping women, and massacring settlers in their homes. The fort lay in the exact center of the state, five miles (8 km) east of Ellsworth, a popular stopping point for travelers on the Santa Fe Trail. The Kansas Pacific Railroad was building west at the rate of two miles (3.2 km) a day, and Ellsworth was the end of the line.

Doctor Sternberg found housing at the fort crude and limited, not suitable for an officer's bride. So Maria reluctantly returned to New York, and her husband went house-hunting. He filed a homestead claim, then bargained with other officers for more property, especially any with improvements. Within a short time he acquired a ranch of approximately 600 acres along the Smoky Hill River, south of the fast-growing town of Ellsworth.[9] A house on property he bought had several rooms and would be adequate for a growing family—and he knew where to find just the right family for his farm.

In the spring of 1866 he wrote to his father in Iowa, asking him to come and look over the ranch. Possibly, he wrote, there was a future for the family in the broad expanse of fertile soil and great pasture land with a tiny river meandering through it. George bought cows and chickens and hired help to operate the place while he envisioned a more stable family operation in the near future.

Levi made the trip to Kansas, and together he and George developed a plan. There would be work for all seven sons, and maybe even a daughter or two! But Levi realized he couldn't count on the daughters in making future plans for the family: one had fallen in love in Iowa; and Levi doubted she would move to Kansas. Levi was committed to teach until the close of school in the spring of 1867. Then, he agreed, he could move the family to Fort Harker. Now George began spending his spare time fixing up the house so it would be fit for Maria, and finally he sent for her to come home to him. She arrived in May 1867. They had been separated for a year, and both eagerly anticipated a second honeymoon at the ranch. But their time alone was shorter than expected—two of George's brothers would arrive soon.

When Levi returned to Iowa, he explained the plan to his family and announced they would move sometime in the summer of 1867. He described land in Kansas to Margaret and the children as vast prairies with gently rolling little hills, very small streams, lots of grass, usually some wind—occasionally very strong wind—and very few people, except at the fort. He didn't know anything about the topography or geology of the state, and it didn't occur to him that such knowledge would be important. He couldn't have known what potential Kansas held for launching the fossil-hunting career of his son Charles—and later, of his three grandsons. For the present, he was content to abide by his teaching contract, and, when he had time, to find out more about this wild state of Kansas, a turbulent region during the war just ended and a gateway to the West.

Charles and Edward, the seventeen-year-old twins, were the most excited at the prospect of moving West. They were full of questions, most of which their father could not answer. Were there still wild Indians? Were there battles? Were there lots of covered wagon trains moving west? What was life really like at a fort? Their imaginations fully engaged, the boys could hardly contain themselves. They struck out by themselves for Kansas early in the spring of 1867, ahead of the rest of the family. They felt no hesitation about going, sure that their brother George would be glad to have their help at the ranch as soon as possible.

When they crossed the Missouri River at Kansas City and boarded the Kansas Pacific train, their enthusiasm was as expansive as the countryside. They had read about "Bleeding Kansas" and John Brown's violence in 1856 and had heard about how, in 1863, William Clark Quantrill and his pro-slavery forces had raided and burned Lawrence, the first town west of Kansas City. The massacre left 150 dead and most of the town destroyed. Now four years later, as the Sternbergs passed through en route to their new home, Lawrence was still picking itself up after the Quantrill devastation. But with the motto "From Ashes to Immortality," the town's attitude was positive, and on Mount Oread, just outside the little city, a state university was rising.

The boys looked intently for charred evidence of the disastrous fire, but actually saw very little as the train moved steadily west.

The train passed through Manhattan, where a land-grant school—later to become Kansas State University, dedicated primarily to the study of agriculture and science—was in its early years. A nice little place, the boys thought; but little did Charles realize that in only a year he would be a student there. Then they went through Fort Riley, the largest and most permanent of all the forts in Kansas. It was exciting, and they wondered: Would Fort Harker be bigger? Smaller? Was it full of soldiers and generals and maybe some captured Indians? Just what could they expect?

They looked at the broad, open spaces. Trees were farther apart with each mile they traveled, and the land was flatter. The boys didn't know much about geology so they couldn't understand the glacial deposits that lay atop the soil of the eastern counties, the various strata of limestone and shale outcropping. The tall grass of the Flint Hills lay thick on the soil after its winter's rest. Finally, they saw the red sandstone of Dakota formation of central Kansas.

Kansas land slopes up to the west, rising an average of ten to fifteen feet a mile (about 3 to 4 m a kilometer) from an altitude of about eight hundred feet (240 m) near the Missouri border at the east to four thousand feet (1,200 m) at the Colorado border, four hundred miles (640 km) west. Several thousand feet—in some places over ten thousand (16,000 km)—of sedimentary rocks, shaley and clayey limestone, sandstone and siltstone lie below the surface of Kansas. The rock strata tilt slightly downward toward the west. Thus, the eroded edges of the oldest formations appear near or at the surface in eastern Kansas at the low end of the incline, and the youngest formations crop out at the surface in the west.

A rock quarry near Hays, Kansas, supplies chalk rock for many uses, but this formation lacks the hardness of post rock, found about 50 miles (80 km) east of Hays. —Ellis County Historical Society

Limestone fence post in Kansas around 1930. —Dr. L. D. Wooster photo, Forsyth
Library, Fort Hays State University, circa 1930s

Charles began watching the different strata more carefully as the
train chugged its way from Salina on its last thirty-mile (48-kilome-
ter) stretch. This portion of the railroad was very new. In fact, it went
only to Ellsworth, just five miles (8 km) west of Fort Harker; it would
be another year before the line reached the Colorado border. Charles
detected, with increasing frequency, a distinct and different rock
stratum in the limestone outcroppings. Known as fence post limestone or post rock, it forms a layer about twelve inches (30 cm) thick
and is found in an area beginning about thirty miles (48 km) west of
Salina, north to the Nebraska line and southwest to Dodge City,
roughly five thousand square miles or three million acres (13,000 sq
km).[10] Prompted by the scarcity of wood, Kansas settlers were just
beginning to use the stone not only for fence posts, but also for houses
and barns. A single limestone fence post weighed approximately five

hundred pounds (225 kg) and could defy wind and fire, lasting a lifetime or longer.

The landscape changed as the brothers approached the Smoky Hill River valley, so named because a persistent haze always hangs above it. Ellsworth and Fort Harker were nestled not far to the north, up the valley floor from the river bed.

Fort Harker guarded the route of the Butterfield Overland Despatch and the Smoky Hill Shortcut to Denver, and was a link in the military trail from Fort Leavenworth to Fort Riley, Fort Dodge, and the Santa Fe Trail; but the railroad was the key to the permanence of the settlement and its hope for the future. Originally called Union Pacific, Eastern Division, the railroad changed its name to the Kansas Pacific, then again to Union Pacific, Kansas Division.

Ellsworth County had a population of 1,185 in 1870, one-third (395) of whom lived at Fort Harker. The population increased seasonally in the early fall with the influx of cowboys and drovers on cattle drives from Texas and the Southwest.[11] Settlers moved slowly to break the sod and raise grain. Most commerce and activity in the county centered around the fort.

If taming the land seemed slow, taming the population took even longer. A railroad officer boasted in St. Louis in June 1867 that "progress on the rail line should be rapid unless the men get frightened away by Indians."[12] His fear was not ungrounded. The iron monster was a threat to the Indians, and they fought its encroachment unceasingly.

In June 1867 Indians killed and scalped five railroad men in two separate attacks within twenty miles (32 km) of Fort Harker. Rail workers took to carrying rifles on the job; and, at least for a time, the Indians were subdued. However, in October Indians fired on a

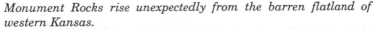

Monument Rocks rise unexpectedly from the barren flatland of western Kansas.

construction train and killed a contractor. Early in the spring of 1868 they strung a telegraph wire across the track in an attempt to stop a train.

A soldier at Fort Ellsworth, later Fort Harker, wrote to his sister in the East: "You would laugh to see the Ft." He went on to describe it as a series of dugouts on a river bank and a few log shanties for the "aristocracy." There were no trees for miles, only stony hills and valleys. Another soldier bemoaned the need to burn buffalo chips to make coffee in the "broad limitless expanse of desert."[13]

Railroad promoters, on the other hand, praised the "brilliant azure" of the skies and weather that "enables you to do more work" there "than elsewhere." Other soldiers and their wives wrote of the varieties of birds and the fragile beauty of native wildflowers. One pioneer woman wrote: "The pleasantest memory I have of pioneer days in Western Kansas is the memory of the prairie in the springtime with the green grass as far as one could see, dotted here and there with millions of beautiful wild flowers."[14]

The Sternberg boys had no conception of what lay ahead for them. But they felt the train slow down. It whistled. Brakes screeched, and the iron monster let off steam and came to a halt. Fort Harker at last! Eagerly the boys piled off the train. There stood Dr. George Sternberg waiting for them. What a wonderful sight! Many months had passed since the brothers had last seen George. In a few minutes they were headed for the ranch, two and a half miles (4 km) south of the fort.

CHAPTER TWO

Kansas Calling

C harles and Edward found a lively and different world at Fort
Harker and in nearby Ellsworth. Settlers moving west—some
by rail, some by wagon train—often stopped at the fort seeking
supplies and the protection afforded by army escorts or the independent scouts who came to the post between assignments to aid travelers
or help squelch trouble with marauding Indians.

The Sternberg boys loved to hang around the fort—especially since
"Wild Bill" Hickok, "Calamity Jane," and "Buffalo Bill" Cody came
there often. Cody, who always had entertaining stories to tell, liked
the Sternberg boys and enjoyed watching their faces when he related
his encounters with Indians. He also told them where they could find
things of interest and encouraged them to explore the countryside.
The twins roamed as far as Fort Zarah, a small post near Great Bend
on the Arkansas River, some forty miles (64 km) southwest of Fort
Harker.

One day, after hearing Cody talk about buffalo—which, he said,
were like birds in that they went north for the summer, then returned
south when winter approached—Charles and Edward hitched a team
of Indian ponies to a light-spring wagon and started southwest to hunt
buffalo. The land was open range for cattle. Not far from Fort Zarah,
the boys sighted buffalo lying down off in the distance. They drew up
the team, climbed out of the wagon, and began crawling to within
shooting distance. Suddenly a rancher appeared, and the buffalo rose
as a mass and quickly loped away. Disappointed, the boys returned to
their wagon and drove on toward the banks of the Arkansas.

When they found paths cut through the deep grass by buffalo, they
stopped, climbed down from the wagon again, and lay in the grass,
waiting. A large animal rushed into view. Charles drew a bead and

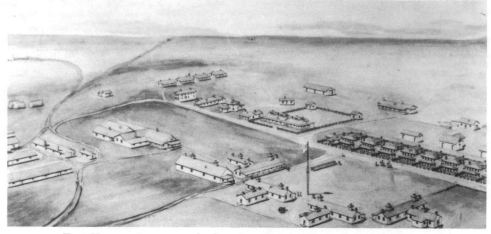

Fort Hays as it appeared when Gen. George M. Sternberg and Gen. George Custer were active. It was a stopover point for Charles H. Sternberg in his early years of fossil hunting, as he worked his way west from Ellsworth and Fort Harker. —Fort Hays State University Photo Service

pulled the trigger. Down went a great brown mass in a heap. "I've killed a buffalo," shouted Charles as he ran toward his prize, only to discover he'd killed a Texas cow. The boys rushed back to the wagon, whipped the ponies into a gallop and headed for home. They convinced themselves the cow was a stray, roaming with the buffalo, and therefore open prey. No rancher appeared to contest their rationalization. On the way home, the boys found an old bull buffalo standing in a ravine—an easy target. They riddled him with bullets, then threw stones at him to make sure he was dead. They continued their trip home, smugly triumphant: they had killed their first buffalo.

Fort Larned and Fort Hays were not far from Fort Harker, and there was considerable interaction between them. Gen. George Custer and some of the famous Seventh Cavalry occasionally had business at Harker, and sometimes there were surprise visitors. One such visitor was Chief Satanta of the Kiowas, who arrived one day in a government ambulance he had captured. After stuffing himself with food and whiskey from the army's stores, he said, "All the property on the Smoky Hill is mine. I want it, and then I want hair [scalps]!"[1]

In the early fall of 1867 chieftains of the Apache, Arapaho, Cheyenne, Comanche, and Kiowa tribes gathered near Medicine Lodge in southern Kansas to sign a peace treaty whereby the United States government would move them to reservations in Oklahoma and limit their hunting and roaming to the unsettled areas south of the

Arkansas River in Kansas. The approaching encampment created excitement at Fort Harker all summer and kept the military on the alert.

While Charles and Edward enjoyed the relative freedom of not having parents around during the first few weeks after their arrival from Iowa, both boys had regular chores and responsibilities on the farm. So it was with a mixture of relief and apprehension that they greeted Levi, Margaret, and the rest of the family when they arrived that summer. There was work for everyone, and Levi quickly assigned duties for each child. Charles's main responsibility was to make the early morning delivery of produce to the fort. Milk, butter, eggs, and garden stuff had to be delivered to officers' homes, as well as to the general kitchen, by 5 a.m. After the delivery, he had time to roam the countryside if he wished.

Charles loved to wander in the red sandstone of the Dakota formation and to study the mushroom-like concretions of the Cretaceous period. Here and there, amid the red, white, and brown strata, layered with various colored clays, he found hundreds of beautiful impressions of fossilized leaves. He took a few home with him, and the next time he visited the area, equipped himself with a collection bag and a pick. Each day, after making his deliveries, he filled his bag with rocks and headed home to show his discoveries to his brother George.

George frequently identified strange old bones found by soldiers, and expressed more than professional interest in the activity of discovering the secrets of the earth in this unspoiled open range. He

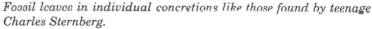

Fossil leaves in individual concretions like those found by teenage Charles Sternberg.

encouraged Charles to handle his leaf imprints carefully and cautioned that they might be valuable. He identified a few specimens for Charles and mentioned that he had friends back East who would know exactly what the fossils were and would have an idea of where they might be sold. Charles carefully laid them aside to wait until the doctor could help him send them off.

The summer that began so well ended in great loss for Dr. George Sternberg. He had brought his family to his ranch and was glad to see them settled and apparently happy. He and Maria moved from the ranch into quarters at the fort, where, for a brief time, they were able to enjoy a semblance of a social life. As recorded in *Life in Custer's Cavalry*, Jennie Barnitz, wife of Col. Albert Barnitz and a friend of the Sternbergs, wrote: "They [the Sternbergs] have five spacious rooms— very handsomely furnished—with china and silver and with excellent servants. The Doctor has a farm near here which he has cultivated, and his table is furnished from it—onions, radishes, green peas, etc."[2]

But the Sternbergs' happiness was short-lived. A great epidemic of cholera swept through the fort, taking so many lives that the Army deemed it best not to report the numbers each day. On a hot summer night, Dr. Sternberg sent this brief report to the surgeon general in Washington: "One of the ladies of the garrison died of cholera on the fifteenth of July." The report did not identify her, and nothing more was made public about the death of Maria Sternberg.[3]

George was devastated. Later in the summer, after the epidemic ended, he requested a leave and returned East. When he next reported for duty, in December 1867, he was ordered to Fort Riley, Kansas, as post surgeon and court martial duty officer. He returned to Fort Harker periodically to visit his family, but was never reassigned there.

A light went out of Charles's life when George left. They had had only a few months together, and now, Charles knew, it would never be quite the same. However, he resolutely returned to the hilly area where he loved to wander and hunt. He no longer cared to shoot buffalo and carried a gun only for protection. His hunting was a peaceful venture.

He found and named a favorite spot Sassafras Hollow because of the countless fossilized sassafras leaves he quarried there.[4] In Ellsworth, Lincoln, and Russell counties, the ancient leaves fell into soft mud and were buried by the thousands in sediments from the freshwater sea that covered the land many centuries ago. As time went on, the accumulated weight above pressed the mud about the leaves and caused the sand to consolidate. The great amount of iron in the heavy vegetation contributed greatly, it is theorized, to the deep red color of the sandstone in the area. Often the iron that accumulated

Charles H. Sternberg returning to his teenage years' favorite spot, sits amid a pile of concretions containing sassafras leaves in Ellsworth County.

about the leaves made the rock very hard. As softer rock formed, then washed away, the hard concretions were left; when broken open, they reveal the impressions of both sides of the ancient leaves themselves having decayed centuries ago.

One morning Charles made his usual delivery to the fort, then turned his wagon toward home. His pocket was full of cash and bills from his sales, and the world looked bright. He decided to go directly home and not stop to hunt more fossils. Suddenly, weariness overtook him, as it sometimes did. Knowing that all he had to do was head the horse in the right direction and the faithful animal would make it home on its own, Charles lay down in the wagon and fell asleep. He never remembered just what happened; but when he reached the ranch, his brothers found him sitting up in the wagon, moaning and incoherent. Blood flowed from a sling-shot wound in his forehead, and he could not hear out of one ear. He had been attacked while asleep and robbed of all his money. He never regained hearing in that ear.

A neighbor summoned the post doctor, B.F. Fryer, who rode to the farm, treated the boy, then contacted George at Fort Riley, approximately ninety miles (144 km) away. When Charles regained consciousness about two weeks later, he saw George lying on a mattress on the floor beside him.[5] A powerful love bound these brothers, and it brought rich benefits to Charles throughout his life.

Charles H. Sternberg once attended classes in the towered building in the background of this 1885 view of Kansas State Agricultural College, Manhattan. —Kansas State Historical Society

The carefree days of youth gave way to adulthood for Charles. He read whatever he could find about fossils, fossil hunting, and the men who were making national news with expeditions and discoveries of fossils in the newly opened West. Levi was pleased with his son's evident interest in reading, but nevertheless discouraged his fossil-leaf hunting expeditions.

Meanwhile, Levi wasted no time after getting his family settled in Kansas. He made his presence felt and his talents available for religious and community work. He preached whenever and wherever he had opportunity, usually for a Presbyterian congregation since there were few Lutherans in the area. He also established contact with educational institutions within a reasonable distance of Fort Harker. He met the Reverend Charles Reynolds, chaplain at the hospital at Fort Riley, and learned that he was a member of the Board of Regents at the newly established Kansas State Agricultural College (now Kansas State University) in Manhattan, and that Reynolds also served on the school's Board of Visitors. Within a year the Reverend Levi Sternberg was also listed as a member of the Board of Visitors.[6] What their duties entailed is not known, but since the school had mandatory attendance for weekly chapel services, it is a safe bet that the ministers on this board assisted with, or conducted, these services.

Probably it was more than coincidence that Charles and a younger brother enrolled for the college's Preparatory Classical curriculum in the fall of 1868. Levi expected his sons to obtain a college education, if possible. At college, the brothers studied geography, arithmetic, grammar, Latin, Greek, and United States history.[7]

The young Sternbergs returned to the farm as soon as classes ended in the spring, and Charles spent more and more time exploring the countryside and hunting fossils. He most commonly found leaf impressions, which he hoarded until he had a vast collection of many varieties. He remembered brother George's offer to help him find an outlet or market for them, and at the first opportunity solicited his help in selecting different varieties and carefully packaging them for shipment to the Smithsonian Institution and to George's medical school friend, Dr. Leo Lesquereux, a Swiss botanist then in the United States.

The Smithsonian did not pay for the fossil leaves, but a letter of acknowledgement from Spencer Baird became a memento cherished by Charles the rest of his life. It read:

Washington, June 8, 1870

We are duly in receipt of your letter of May 28th, announcing the transmission of the fossil plants collected by your brother and yourself and shall look forward with much interest to their arrival. As soon as possible after they reach us, we shall submit them to competent scientific investigation and report to you the result.

Very respectfully yours etc.,
Spencer F. Baird,
Assistant Secretary in Charge[8]

It was several years before the specimens were finally examined and the findings were published by Dr. John Newberry, professor at Columbia University and state geologist of Ohio. By then, the name of Sternberg as collector had been forgotten or overlooked. Consequently, Charles received no credit. Some years later he recognized plates of specimens he had collected in Newberry's *Later Flora of North America*.

George remarried in 1869 while stationed at Fort Riley; and he and his wife Martha stayed in touch with Charles and his fossil hunting aspirations as closely as possible. The fame of the doctor's interest in fossils and strange, old bones reached from one fort to another; and one day he received a collection of fossils from Dr. Theophilus Turner at Fort Wallace, near the Kansas-Colorado border. Sternberg bided his time until he had occasion to go to Washington, then took the fossils to paleontologists in Philadelphia. Their reports provided the first record of fossils found in extreme western Kansas and opened the door for teams of fossil hunters to search the Kansas plains with intensity.[9] The doctor also sent geologic specimens to Dr. Joseph Leidy of Philadelphia who mentioned them in his report to the Army Medical Museum on "The Vertebrates of the West."[10]

As Charles learned about his brother's fossil-oriented activities, his enthusiasm increased for what he had decided was to be his life's work—and he told his family that he intended to devote his life, whatever it might cost him in "privation, danger, and solitude," to make it his business "to collect facts from the crust of the earth" so that "men might learn more of the introduction and succession of life on our earth."[11]

George fully supported Charles's resolve, but the rest of the family did not share his enthusiasm on the grounds that such an ambition was not practical for a poor man's son who would always have to work for a living. In the summer of 1870 George was assigned to duty at Governor's Island, New York Harbor. His life in Kansas was over. But for Charles, George's reassignment was simply one more challenge. He would pursue his goal alone, but with assurance of support from George, wherever he might be sent.

Charles continued his wanderings in the hills and valleys near Ellsworth, returning frequently to Sassafras Hollow, which lay at the head of a narrow ravine in a ledge of sandstone, with a spring beneath. Not far in another direction, he explored a sheer claystone bluff southeast of Fort Harker and found two of the largest leaves known at that time to have come from the Dakota formation, each more than a foot (30 cm) in diameter. One, with three lobes and a stem that passed through an ear-like projection at its base, Dr. Lesquereux called *Aspidophyllum trilobatum*. The other, also with three lobes, but indented with what appeared to be teeth, Lesquereux called *Sassafras dissectum*.[12]

Dr. Lesquereux visited the area in 1872 and among other specimens found a large, beautiful leaf that he named *Protophyllum sternbergii*. Charles was overwhelmed at this first scientific honor; consequently, this particular leaf always stood out in his memory as his first claim to fame.[13]

During his visit, Lesquereux was a guest of Fort Harker's commander, who gave a reception in his honor. When Charles learned of the event, he tucked under his arm several sketches he had made of the first specimens he had sent to the Smithsonian and headed for the fort. Lesquereux was deaf and spoke only French, but he was accompanied by his son, who served as interpreter. As soon as introductions were over, Sternberg pulled out his drawings, and a lively conversation ensued. Moving off to a corner of the room, the famed botanist was obviously impressed by what he saw. As Sternberg later told the story: "His eyes shone when he examined the drawings. 'This is a new species, and this, and this. Here is one described and illustrated from poorer material.' "[14]

Encouraged by the attention and recognition, Charles began sending Lesquereux all of his collection for description. He had more than four hundred species of plants like those growing in existing forests along the Gulf of Mexico—some vines, ferns, and even the fruit of a fig and a magnolia flower petal—all found in the Cretaceous of Kansas![15]

Charles established a lively correspondence with Lesquereux that continued until the death of the botanist in 1889. He also continued to clean, smooth, and polish his leaf specimens. When he felt he had done all that was possible without injuring the imprints, he laid them away on trays to be numbered and identified.

The summer of 1872 passed swiftly for Charles. He enjoyed his outdoor work, now mostly herding cattle, and as the days grew shorter and the wind sharper, he bundled himself into warmer coats and continued to spend every possible moment searching the hills and rocks—all the time keeping an eye on the cattle.

One day, late in the fall, a furious sleet storm developed. Charles was far from the ranch and shelter, having taken the cattle some distance farther than usual to find water. The stream was frozen and, in cutting the ice, Charles splashed the frigid water over much of his body. His legs were thoroughly soaked; by the time he reached home his clothing had become a solid cake of ice, and his bad leg felt stiff and numb. The chilling of his bones left him somewhat incapacitated, and he spent much of the winter in a big chair close to the stove while his mother tended to his needs. When the inflammation finally subsided, Charles's left knee was stiff. To avoid having to spend the rest of his life on crutches, he went to the hospital at Fort Riley, where, through George's influence, he was given special care. He spent three months in the hospital and underwent surgery so that he was finally able to throw away his crutches and cane. Nevertheless, his knee remained stiff for the rest of his life. He observed that he had "walked thousands of miles among the fossiliferous beds in the desolate fields of the West, always with a limp."[16]

The army chaplain, Rev. Charles Reynolds, visited Charles frequently in the hospital and sometimes brought his daughter Anna with him. Rev. Reynolds was a friend of George, as well as of Charles's father, with whom he had worked on the Board of Visitors at the college in Manhattan. Charles looked forward to the visits of the Reynoldses and noted the beauty and grace of Anna, whom he assumed to be about his own age—twenty-two. However, he had no particular interest in, nor time for, romance at this point in his life.

Recuperation was long and slow for Charles. But for the Sternberg family, times were changing. Fort Harker closed in 1873, and, despite the efforts of Levi and others, the land was sold to private developers. In anticipation of such action, Levi had made appeals to federal and state officials and the citizenry in general to set aside the lands for educational and public purposes, such as forest and fruit culture. Always ready for a soapbox oration, sermon, or editorial, he had over-optimistically predicted there would be no serious obstacle in securing "such disposition of Fort Harker to the action" of the "state legislature and ... Congress. The buildings are all ready for occupancy, and the sale of such portions that would not be required for the purpose indicated would furnish the endowment for an institution of learning which would, thus, like Minerva, leap fully armed, from the cloven head of Jupiter."[17]

Levi had an ally—and competitor—in Hays (City), one Martin Allen, who had similar ideas for the use of the Fort Hays land should that fort also be abandoned. The two men were friends with common interests in horticulture, farming, politics, and writing, but each saw distinct advantage in preserving the fort lands. Sternberg had given

Levi Sternberg preached at the first service held in the first church erected in Hays, Kansas, 1879. Had he looked, he could have seen fossils imbedded in many of the rocks of the structure. All materials for the church were quarried from native stone, rich in fossil deposits.
—Ellis County Historical Society

26

up trying to find enough Lutherans to establish a congregation and had attended a seminary to accredit himself as a Presbyterian minister. He visited Allen in Hays, and delivered the sermon at the opening of a Presbyterian church built in 1879. It was a stone building, erected from locally quarried limestone, and if Sternberg had examined the stone blocks, he could have seen countless fossils imbedded in each one.[18] But his mind was not on fossils.

The efforts to save Fort Harker lands for public use were futile, and the little community soon disappeared. Levi and several of his married sons were among the many residents of the fort and surrounding farm area to move to Ellsworth.

By the fall of 1875 Charles had regained his health and done a lot of reading and thinking that led to renewed interest in college; so he enrolled as special student at the Manhattan school. One of his professors, B.F. Mudge, was organizing a fossil-hunting expedition to set out as soon as the weather opened up in the spring, under the auspices of O.C. Marsh, who was associated with Yale College (now Yale University). Charles applied for inclusion in the group, but enrollment had already closed. Keenly disappointed, he sought other opportunities, and in desperation wrote to Edward D. Cope in Philadelphia:

> I put my soul into the letter I wrote him, for this was my last chance. I told him of my love for science, and of my earnest longing to enter the chalk of western Kansas and make a collection of its wonderful fossils, no matter what it might cost me in discomfort and danger. I said, however, that I was too poor to go at my own expense, and asked him to send me three hundred dollars to buy a team of ponies, a wagon, and a camp outfit, and to hire a cook and driver. I sent no recommendation from well-known men as to my honesty or executive ability, mentioning only my work in the Dakota Group [formation].
>
> I was in a terrible state of suspense when I had dispatched the letter, but fortunately, the Professor responded promptly and when I opened the envelope, a draft for three hundred dollars fell at my feet.
>
> That letter bound me to Cope for four long years.[19]

School days, herding cattle, and boyhood were things of the past. Charles Sternberg was ready to move out and begin his life as a fossil hunter.

CHAPTER THREE

Indians and Fossils

Young Sternberg learned many things that winter of 1875-76 in Manhattan, among them that fossil hunting was a real and organized business dominated by two men—Othniel C. Marsh of Yale University and Edward Drinker Cope of Philadelphia. Charles had heard of these men through his brother George, but he still had much to learn about them and the complexity of their profession. Marsh and Cope were onetime friends who became intense and bitter rivals, jealous and suspicious of each other. Each employed teams of workers to do most of the actual fossil hunting for them.

As the first professor of paleontology in the United States, Marsh played a major role in establishing the study of vertebrate fossils in American universities. His main interest lay in fossil horses and other early mammals, toothed birds, and dinosaurs.

Born in Massachusetts in 1831, he came from a humble family, but his maternal uncle was George Peabody, an English financier. Following his mother's death, this rich uncle took young Marsh under his wing and educated him at Phillips Academy (Andover) and then at Yale. At Marsh's suggestion, Peabody established the Peabody Museum of Natural History at Yale and provided a non-teaching professorship for him there.[1]

As a youngster, Marsh had dallied in school and delayed planning his own life until he was well past the usual age for making such decisions. He finally decided to become a scholar in order to win the approval and support of his wealthy uncle. Suddenly life became a challenge. Marsh developed a great drive and passion for achievement and recognition. His ability as a manipulator led him into politics where he used his charm and knowledge to achieve his goals. His strong will and forceful opinions became paramount in his character.[2]

Edward D. Cope.
—Fort Hays State Photo Service,
copied from *Life of a Fossil Hunter*

In contrast to Marsh and his modest beginnings, Cope was born to a wealthy Quaker family in 1840 near Philadelphia. While a child, he became obsessed with natural history and writing. At age six he wrote sophisticated notes concerning the skeleton of an ichthyosaur, a marine reptile that he saw in the Philadelphia Academy of Sciences. Cope turned his back on the family's farming interests, instead devoting his time to research and writing. At age eighteen, he had published his first scientific paper and won general recognition as a scientific writer. By twenty he was a researcher for the Smithsonian.[3]

But Cope's fervency interfered with attempts to reconcile his knowledge of science with his family's devotion to Quakerism; he became self-righteous, narrow-minded, and intolerant. After the death of his father, he became increasingly generous and indulgent as he pondered Darwin's theory of evolution.

Cope and Marsh were both in Europe in 1863 as young men, studying and observing at major museums. They became casual friends. Back in the United States in 1868, they hiked amicably and hunted fossil reptiles together in the Cretaceous marls near Princeton, New Jersey. They were reasonably successful, and Marsh even honored Cope by naming one giant marine lizard or mosasaur specimen *Mosasaurus copeanus* (now known to be *Plioplateocarpus depressus*).[4]

The two paleontologists developed vastly different personal lifestyles. Marsh became more and more self-centered and status conscious. He never married and in his mature years lived a pleasant life in New Haven, near Yale, in a large mansion where he entertained lavishly. Cope, in contrast, married his cousin, Annie Pim; together with their daughter, Julia, they moved to Haddonfield, New Jersey, just east of Philadelphia, where Cope happily spent much of his time writing. His bibliography includes about fourteen hundred published works.

Cope had studied in Berlin, Paris, and London, intriguing museum curators and diagnosticians. When he returned to the United States, he became a zoology professor at Haverford, a Quaker college in suburban Philadelphia. However, he resigned after three years to study paleontological excavations and mineral veins in northwest Nebraska, Wyoming, Ohio, and North Carolina.

Both Marsh and Cope were determined, talented, and ambitious. Science and industry profited from their skills. However, their differences and animosities brought forth power struggles that made frontpage headlines in major New York newspapers and altered the national search for fossils in the final quarter of the nineteenth century. One incident that might have triggered the feud occurred when Cope erred in the reconstruction of an *Elasmosaurus*, a giant plesiosaur. He actually placed the head on the tail of his restoration, a blunder that Marsh took advantage of and forever humiliated Cope. Cope worked in haste to repent at leisure.[5]

Visiting the Smithsonian, Marsh learned of the discovery of fossil leaves in Ellsworth County, Kansas; and through his acquaintance with Dr. George Sternberg, he realized there were many more fossils where these leaves came from. In 1870 Marsh set out to explore the territory for himself. He took twelve Yale students with him to study the strata and fossil beds from Nebraska to the Pacific coast.

Marsh prepared his men for the hardships he knew they would have to endure. He required them to read and study *The Prairie Traveler*, which taught them how to make camp, build temporary boats, start backfires to halt prairie fires, deal with Indians, and cope with other perils. Each man was well equipped with rock hammers and waterproof boxes of matches and carried a bowie knife, a 50-caliber Sharp's carbine and a 36-caliber Smith and Wesson six-shooter. Each had sufficient cash to meet his needs.[6]

Marsh always arranged military escorts for his expeditions. "Buffalo Bill" Cody accompanied his party for at least a day, leading to a lifelong friendship between Marsh and the scout.

Marsh also won favor with landowners, and with great foresight and efficiency made arrangements with them to save material for him and to exclude other hunters wherever he sought fossils. Both Marsh and Cope taught their men how to maintain secrecy and how to hide the exposed bones and protect the area from rival hunters.

Marsh and his students traveled by wagons across Nebraska and left them in Cheyenne, Wyoming. They went by train across the Laramie Plain to Fort Bridger and into Utah where they searched the huge boneyards that extend from the Uinta Mountains to the junction of the Green and White rivers. They found no large dinosaur bones, although they were firmly convinced they were there somewhere, just waiting to be discovered. The group returned to Cheyenne and from there to Denver and southeast into the Smoky Hill River valley of Kansas. After hunting there, they headed for home.

Reports of Marsh's expedition with his Yale students provided much discussion for Charles Sternberg and fellow students at Kansas State Agricultural College. They also discussed the experiences and contributions of Joseph Leidy, Ferdinand Hayden, Samuel Williston, and other paleontologists who dominated the period from the Civil War to the end of the nineteenth century. Charles's brother George knew several of these men and had told Charles about their explorations and discoveries.

Ferdinand Hayden was a medical doctor who, after serving in the Civil War, turned explorer and geologist. As early as 1856, he was in the Judith River basin of Montana and sent Leidy fragments of fossilized dinosaurs. He also found the first real dinosaur skeleton in North America—*Hadrosaurus foulkei*, a duck-billed vegetarian discovered near Haddonfield, New Jersey, not far from Cope's home.[7] Hayden shipped many fossils to the Smithsonian, further convincing scientists that the wilderness west of Missouri River contained the most significant array of fossils ever discovered.

Cope went to Kansas in 1871 with seven men, two wagons, and fourteen mules. He knew Marsh and his Yale group had already been through the territory, but felt there were still fossils to be found. He expected to find flying reptiles, but failed to uncover any trace of them. However, he did fill two wagons with skeletons of sea turtles and an 800-pound (360-kilogram) bulldog tarpon, *Portheus* (also known as *Xiphactinus*), and a mosasaur, *Tylosaurus*. He bragged that he had secured "a large proportion of the extinct vertebrate species of Kansas" even though Marsh had been there before.

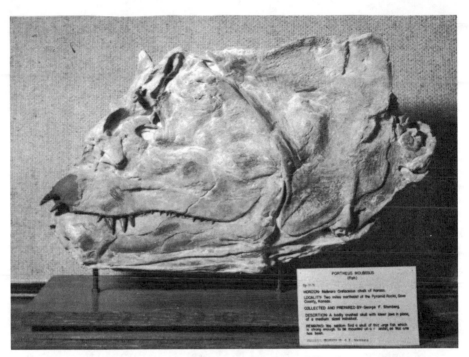

A damaged skull of a typical, medium-sized Portheus molossus (Xiphactinus).

Joseph Leidy made a trip in 1872 with Cope to the Fort Bridger area of Wyoming. Marsh was in the same area. Leidy tried to reconcile Marsh and Cope, and finally each promised to mail to the other copies of "any papers on the subject of fossils from Wyoming, Eocene strata," they might issue, with "the date of publication to be written or printed, on each pamphlet."[8] The rivalry and bitterness between Cope and Marsh increased until Leidy finally withdrew from vertebrate paleontology; his withdrawal did nothing, however, to mitigate the problem.

Marsh was adamant in claiming monopoly on searching rights; it was, he contended, *his area*. Like all fossil hunters, he wanted credit for his finds and wanted to be the first to discover and name new creatures. Fossil hunters received credit only when they published their finds. Irregularities between discoveries and printed reports that credited the wrong person were not unheard of. Recognition was more desirable than cash.

The race to find fossils can be likened in some respects to the gold rush. The rivalry was intense; the scheming and conniving were beyond description; and the jealousy endured for the rest of Marsh and Cope's lives. Few men who worked with them managed to avoid the clash of personalities and the flare of tempers, but Sternberg tried desperately to remain outside of the conflict. For gold seekers, the prize was wealth; for the fossil hunters, the reward was the recognition and honor of having species of fossils named in their honor.

Samuel Williston, like Sternberg, was younger than Cope or Marsh and just beginning his career. Also like Sternberg, he was a free-lancer, willing to work for either Cope or Marsh, but preferring not to be involved with their differences.

Sternberg pondered the reports of the strife between Marsh and Cope and struggled with his own position as he prepared to enter this highly competitive field. Where should he pin his allegiance? Cope's bank draft helped him make up his mind.

His dismay at not being included in the Mudge party setting out from the agriculture college gave way to high spirits and confidence as Sternberg, now twenty-six years old, set forth on his first assignment in the spring of 1876. He was confident Cope would be his friend and mentor and glad he was going to work for him. The next four years proved to Sternberg, at least, that he had made a wise decision. His admiration for Cope became almost worship, and he never faltered in loyalty to his mentor.

Sternberg regarded Cope as the greatest naturalist the United States ever produced and said it was one of the greatest joys of his life to have known and worked with such a man. Probably he would have felt the same way about any other leader under whose sponsorship he worked, so consuming was his love for fossil hunting, despite its hardships, privations and low income. For his part, Cope never made Sternberg feel inferior because of his lack of formal education. Instead, he lauded every accomplishment and contribution, cementing a bond of mutual admiration and friendship.[9]

With the $300 Cope advanced for Sternberg's first expedition that summer of 1876, Charles acquired a team of ponies and a driver, packed the basic essential camp equipment and tools, and headed for Buffalo Park, a station on the Kansas Pacific, about 125 miles (200 km) west of Ellsworth. Buffalo Park consisted of a lone house, occupied by the section men, and a railroad station where Sternberg's brother William served as station agent. The station's most significant attributes were a windmill and well of pure water, 120 feet (36 m) deep: "It made a Mecca for us fossil hunters after two weeks of strong alkali water," wrote Sternberg later. "At this well Professor Mudge's party and my own used to meet in peace after our fierce rivalry in the field as collectors for our respective paleontologists, Marsh and Cope."[10]

All fossil hunting parties had similar basic supplies and equipment—flour, bacon, salt, and beans, as well as cigar boxes and flour sacks to hold small fossil parts. The workers created special crates packed with grass or straw for heavy specimens. Drinking water was often harder to find than fossils. Once a fossil-hunting party located a source, like the one at Buffalo Park, the men would establish a semi-permanent base camp nearby, and then fan out from there for two to four weeks with one man returning to the base about once a week to get mail and to ship crates or specimens.

Sternberg worked over all of the exposures of Niobrara chalk from the mouth of Hackberry Creek in eastern Gove County to Fort Wallace, nearly a hundred miles (160 km) west. There were no trails. He headed south, crossing Hackberry Creek about fifteen miles (24

The Cobra is one of the more spectacular images near Castle Rock in Gove County, Kansas.

km) from the railroad and found a spring of pure water—a rare and precious commodity. He developed another route that led to where the tiny town of Gove now stands. From here he looked west toward Monument Rocks—a strange group of pinnacles that a good imagination could turn into castles, turrets, or human likenesses. Beyond lay the remains of an old, one-company outpost on the Smoky Hill Trail.

The fossil hunters' tents and wagon sheets were of heavy brown duck fabric that blended into the dry, brown buffalo grass and served as camouflage protection from marauding Indians. Sternberg did not fear Indians and carried no rifle when on foot. He said he was looking for fossils, not Indians; and he had no confrontations of any consequence.[11]

Once, after a gentle rain had washed the air and taken the glare from the chalk cliff, Sternberg started down a canyon, scanning the rocks for freshly revealed bits of fossil. Instead he saw a deeply cut fresh pony trail. He followed it back to the river and found that a band of Indians had sought shelter from the rain in a willow thicket, tying twigs together and throwing skins over them to shed the water. Under these shelters the band had cooked and rested, leaving live coals in the ash heaps. They had tied their ponies to the bushes without room to graze. Sternberg concluded there were no women nor children in the party because, as he wrote later, "it makes no difference whether women are white or red, they always lose some of their belongings wherever they go, and there was none of such property at this camp."[12]

Later he learned from scouts that the camp he had come upon had been made by Kiowas, Cheyennes, and Arapahos, who, under Chief Crazy Horse, had headed north to join Sitting Bull in Montana in the fateful battle of Little Big Horn.

Burning with enthusiasm for doing what he really wanted to do and determined to justify the trust he believed Cope had in him, Sternberg claimed (and no one could prove otherwise) that he went over every inch of the acres of exposed Niobrara chalk in Gove and Logan counties that summer. He hoped each moment to find a great and perfect fossil specimen to highlight the mass of lesser discoveries he had so carefully claimed, packed and shipped East.

Only the insatiable drive to succeed enabled him to endure the hardships he encountered. It was obvious early in the expedition that everyone suffered more from lack of good drinking water than from all other ills combined. Recalling the hardships of a single day of that first summer of hunting, he wrote:

> The day is so hot that perspiration flows from every pore; a howling south wind rises and fills our eyes with clouds of pure lime dust, inflaming them almost beyond human endurance. Still no water. The driver, with horses famishing for it, makes frantic

Much of the western Kansas terrain was flat and desolate.

gestures to me to hurry. To ease my parched lips and swelling tongue, I roll a pebble around in my mouth, or, if the season is propitious, allay my thirst with the acid juice of a red berry that grows in the ravines...After hours of search, I find in moist ground the borings of crawfishes; with line and sinker I measure the depth to water, a couple of feet below in these miniature wells. The welcome signal is given to Will [the driver] and he digs a well, so that both men and beast may be supplied.[13]

Except for the few springs and the wells they dug, the only water they could find was alkaline, which had the same purging effect on the body as a solution of Epsom salts. Failing to take his own well-being into consideration, Sternberg worked far beyond reasonable limits and finally fell victim to the ague.

One day while Sternberg was shaking with the ague and near exhaustion, he pushed himself still harder, hoping to find a major specimen to send to Cope, and hit pay dirt with the discovery of a mosasaur skeleton. Its head was encircled by the spinal column and its four paddles stretched out on either side. Disintegrated chalk lay as dust several inches thick over the whole mass. Personal discomfort and exhaustion faded into insignificance. Sternberg leaped into air,

A typical mosasaur from the Niobrara chalk of western Kansas.

his arms extended toward the heaven, and shouted for all the surrounding rocks to hear: "Thank God! Thank God!" Ever the professional, he then carefully brushed away the chalk dust and fully revealed the skeleton of the great reptile.

With his helper, he packed the specimen with excelsior and grass; and on the tortuous 25-mile (40-kilometer) trip to Buffalo Park station he sat in the bottom of the wagon, protecting his discovery. Another attack of the ague brought a headache so bad he thought his head would burst, but, he wrote later, he cared little, as long as he got his "beloved fossil to Professor Cope."[14]

Cope named it *Clidastes tortor* because an additional set of articulations in the backbone enabled it to coil. Because it appeared to Cope to be a veritable serpent, he placed it in his new suborder Pythonomorpha.[15]

One harrowing experience followed another, it seemed to Sternberg that summer, but his determination to prove himself to family and Cope kept him going. In midsummer, after searching for two weeks with no success, Charles and his helper pitched their tent in a canyon near Monument Rocks and shortly thereafter found two specimens of *Platycarpus* lying in a low knoll, separated by only three feet (1 m) of chalk. No sooner had they begun to clear the fossil area when the weather changed abruptly and a cold rain fell. Fuel was mostly buffalo chips, but fortunately the men found remains of a dead cottonwood tree that they lugged to camp and burned intermittently for the next three days while the storm continued. Their food supply gave out, with little remaining except the corn originally intended for the ponies. Sternberg ordered the cook to parch a quantity of the ponies' corn; the men ate all they could and then stuffed their pockets with it and returned to work.

The strenuous handwork took its toll on Sternberg's palm where the continuous use of a large butcher knife to cut away chalk from the specimen caused a felon to form. A fistula developed and Sternberg was forced to take ten days off to let it heal. He rested at Buffalo Park

and asked Cope for help. Cope sent J.C. Isaac from Illges Ranch, Wyoming, but even the arrival of a fresh companion didn't lighten Sternberg's malaise. Isaac, at best chronically afraid of Indians, was reeling from having seen five of his Wyoming associates shot and scalped by a marauding band. He fancied he saw an Indian behind every bush or rock; and it wasn't long before Sternberg also had Indian jitters.[16]

It was only July, but Cope urged Sternberg to go home to rest and recuperate. Cope had arranged an expedition to the Judith Basin and badlands of Montana for late summer and wanted Sternberg and Isaac, if possible, to join him in the adventure

Isaac and Sternberg returned to Charles's parents' home in Ellsworth where his mother tended his infected hand and nursed him back to health. When Cope sent word a few weeks later for Charles and Isaac, they packed their gear and headed to meet him in Omaha. Sternberg's adrenalin surged; he could hardly wait: at last he would meet the great man who had given him his chance to be a real fossil hunter.

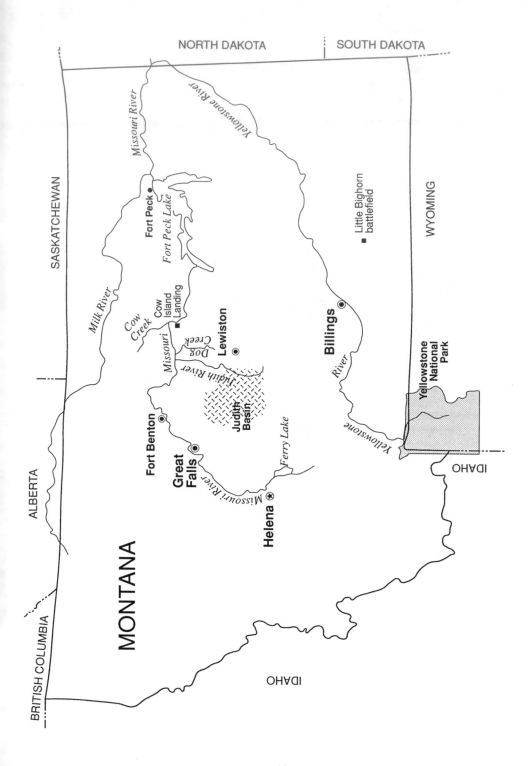

CHAPTER FOUR

Dinosaurs and Rice

I t would be hard to say who was the more shocked when Sternberg and Cope met in Omaha. Cope, who was delicate and frail as a child, never became robust. However, he lived intensely and drove himself with a passion for science and all the mysteries he believed could be unraveled through the study of the fossil remains of ancient animal life. He never doubted his own strength and worked long days with a frenzy he could not explain, but accepted as the normal way to live and work.

As Sternberg stepped off the train he saw a frail man who appeared older than his thirty-five years. Cope swayed slightly, as if unable to resist the summer wind. His wife, Annie, stood close by, ready to assist him if necessary; but with dogged determination he straightened his posture and, with only a slight stagger, stepped forward to greet Sternberg. Cope's regimen of research and writing—he had just finished a fourteen-hour day of work on a Hayden publication—left him ill-prepared for the rugged outdoor life he was facing in the weeks ahead.

Sternberg and Isaac moved closer to Cope, watching his constant swaying and signs of insecurity. Sternberg wondered: Could this man ride a horse? Could he lead an expedition through difficult terrain?

Cope, on the other hand, saw in Sternberg a sinewy, bronzed 26-year-old man, hardened by sun and wind, but so stiff-legged from his childhood injury that Cope, too, was skeptical. He muttered to Isaac, who stood close by, "Can this man really mount a horse?" Isaac said: "I've seen him mount a pony bareback and cut out one of his mares from a herd of wild horses."[1]

They left Omaha by train for Ogden, Utah, where Annie Cope reluctantly bid her husband farewell and returned to the East. Cope, Isaac, and Sternberg continued to Franklin, Idaho, where they took a

Concord coach for the 600-mile (960-kilometer) trip north to Helena, Montana. Bouncing along at a ten-mile-an-hour (16-kilometer-an-hour) pace, fighting dust so thick that even the horses were invisible from the coach, they traveled day and night for three and a half days with stops only to eat and change horses and drivers. The coach bounced over chuck holes and rough terrain, until the passengers feared for their sanity, if not their lives. At one point Sternberg cradled Cope's head in his arms so he could nap. The drivers were less than reassuring. One, known only as "Whiskey Jack" and so drunk he could barely stand, insisted on racing at reckless speed, even in the dark of night. Highwaymen frequently threatened to waylay the coach. Cope later recalled that he finally put "some loose $120 in the facing of his pantaloon legs—a process called 'banking.'"[2]

Rumors of Indian attacks and the news of the Custer massacre greeted the fossil hunters' arrival in Helena. Only six weeks earlier, on June 25, 1876, Sitting Bull and company had wiped out George Custer and his Seventh Cavalry about 150 miles (240 km) from the Judith River basin where Cope and Sternberg were headed. The basin was considered neutral ground between the Sioux and the Crows—a good place for either tribe to kill whites and blame it on the other.

Cope brushed aside all warnings, maintaining that it was an excellent time to be there and that all Sioux warriors were with Sitting Bull. There would be no danger until the United States Army drove them back north. Cope believed he had three months of safe hunting time before the winter's icy blasts would drive the fossil hunters back home. Sternberg was so excited about the whole expedition he could offer no complaint, and poor Indian-shy Isaac merely shivered in fear and stayed with the crew.[3]

From Helena to the Judith basin, they faced another perilous journey by stagecoach. A broken wheel forced a shift to an open wagon with no springs. Rain had washed out the trail and bridges; and one night a driver became lost in the mountains. Eventually they reached Fort Benton where Cope bought horses—four to pull the wagon and three ponies to ride—and supplies; he also hired a cook, and a part-Indian scout named Jim Deer. After packing the wagon, the little expedition headed southeast toward the mouth of the Judith River.

Sternberg, never known for his love of animals, now needed to prove his ability to handle the bad-tempered mustang pony he rode, as well as to help control the undisciplined horses that pulled the wagon. "With two worn out mustangs and a balky colt," Sternberg wrote later, "we were obliged to punish them constantly to keep at work. One of the leaders, a fine four-year-old colt, had to be knocked down half a dozen times before he could be taught not to balk and strike out with his forefeet at everybody within reach. The other leader, Old Major, as true as steel, often saved the day doing his duty nobly in spite of miserable company." Sternberg also wrote about the curb bit that "almost tore his [pony's] mouth to pieces" but was his "only means of restraining him."[4]

Isaac's fear of Indians was not groundless. A tepee village of more than two thousand Crows lay across the river from the Cope camp. Isaac insisted that a guard be on duty all night and offered to take his turn. Sternberg took the first watch and had just returned to his tent when he heard Isaac, his replacement, shout, "Halt!" Cope and Sternberg went to investigate and saw an Indian and his squaw approaching with Isaac leveling his Winchester at them and repeating his demand that they halt. The Indian pleaded, "Me good Indian. Me good!" Cope came quickly to Isaac's assistance and managed to placate the Indians and quiet Isaac's fears. Using a few words, along with some sign language, Cope made the Indians feel welcome and invited them to come for breakfast and to bring along their chiefs.[5]

Cope was cleaning his dentures as the Indians approached the camp that morning; and the visitors watched in silent awe as the white man slipped his teeth in place and advanced toward his guests with a broad smile. "Do it again. Do it again," they begged. Cope smiled and obliged, again and again. For days thereafter, the Crows gathered to watch the morning ritual performed by "Magic Tooth."[6]

There was no more fear of thievery or attack. Instead, the Indians brought gifts of sage hens, fresh fish, and mountain sheep. Cope and Sternberg roamed over the meadow, finding a few fossils, but also making friends with the natives, which was equally important at that point. The pleasant social interlude ended, however, when the Cope party moved to Dog Creek, near the junction of the Judith and Missouri rivers. The narrow Dog Creek valley was nestled below peaks, pinnacles and ridges, some reaching a height of twelve hundred feet (360 m) above the valley floor.[7] The terrain of the Judith basin was a constant challenge, but hopes remained high in the hearts of the fossil hunters. They were finding enough scattered bones and teeth of dinosaurs to whet their desire and imagination. Here had roamed Cretaceous giants—some similar to, though much larger than, the New Jersey iguanodonts and hadrosaurs Sternberg had heard about before he moved to Kansas, and others never before encountered. Here lay an unexplored graveyard of a variety of dinosaurs. It was Charles Sternberg's first exposure to the great prehistoric reptiles.

The men worked with feverish energy. Cope, unable to control his scientific mind, would burst into lengthy lectures on the past, combining the theories of his inborn Quaker training with his readings of Darwin. Sternberg listened in awe, absorbing every word from "the

Cope and Sternberg roamed the rocky formations near the Missouri and Judith rivers above Cow Island.

master naturalist [Cope] who saw beauty even in lizards and snakes" and believed and inspired the belief "that it is a crime to destroy life wantonly, any life," as Sternberg wrote later.[8]

Cope appreciated Sternberg's religious background and beliefs. For too long Cope had suppressed his own Quaker creationist beliefs, and now, at last, he had a sympathetic ear to hear his outpourings. Sternberg hung on every word of scientific theory and tried to reconcile these theories with his fundamentalist beliefs in the Biblical account of the Creation.[9]

Even in sleep, Cope's mind worked full tilt. In his nightmares, horned creatures twice the size of hippos trampled him and dinosaurs threw him in the air and kicked him. He thrashed about until his groans and outcries wakened Sternberg who would soothe Cope. Then they would lie down again, only to repeat the horrendous routine.

Fears of Indian raids still occupied the minds of the cook, Isaac, and Jim Deer. When Deer heard from a scout that Sitting Bull's camp was only a day's march away, he and the cook fled, leaving an extra burden on Isaac and Sternberg. "We willingly undertook the task [of cooking], even if it were to mean working our fingers to the bone," Sternberg said.[10]

Although slowed down somewhat by the decreased work force, Cope and Sternberg were determined to continue their plans. They camped near the mouth of Dog Creek in an open valley between two ridges towering approximately twelve hundred feet (360 m) above them. The horses continued to present problems, yielding only after Cope subdued them with repeated beatings. The trip to the campsite was arduous, with swarms of gnats clustering about the men, getting under their hat rims and inside their clothing, and leaving behind sores with heavy scabs. The horses, too, suffered from the tiny pests, and the only relief came when the men used cold bacon grease under the horses' saddles and collars and on their own faces and arms.

Offsetting the physical discomfort was the exhilaration of finding the skull of a new specimen—a horned dinosaur, *Monoclonius*, characterized by large heads with a bony frill resembling horns. Both Cope and Sternberg found examples of *Monoclonius*, but only in fragments and scattered bones.[11]

Protecting and preserving the fossils after excavation was a continuing challenge for all fossil hunters. Grass or straw packed around the loose pieces of bone or rock sections was far from adequate protection. Many of the bones had mineralized into agate, jasper, and other brittle compounds, and were easily damaged. Now faced with the necessity of devising a glue or other means of holding the *Monoclonius* fragments together, Sternberg improvised a solution—inspired by a sack of rice resting near his feet. Neither he nor Cope really liked rice: they'd had too much of it; and the sack they'd brought was probably wormy anyway, they surmised. But when the rice was cooked, it became thick and gooey and would harden like glue. They cooked the sack of rice, tore burlap and other cloth into strips that they dipped into it as it cooled. Then they bound the strips around the skull of the *Monoclonius* and other fragile specimens until they resembled bandaged broken bones. By morning the mass was hard, and the precious bones well protected. The heavy bundle lay in the wagon, an anchor for smaller bundles and fragments packed with straw and grass.[12]

Over time, they improved the rice-paste technique by substituting flour paste or plaster of Paris for the rice glue. They found that plaster of Paris poured over the surface of a bone while it was still in the rock would protect any delicate projections from injury during the rest of the extraction process. Marsh and his men developed a similar technique a little later. Both Marsh and Cope claimed first right of discovery and jealously guarded their "secret formulas."

The days grew shorter, and the wind had a winter chill. It was October, time to break camp and take the load overland to Cow Island in the Missouri River, forty miles (64 km) away, to catch the last boat of the season to Omaha. The prairie was twelve hundred feet (360 m) above the campsite, and raising the wagon up the steep slope required all the ingenuity the men could muster.

Cope had gone ahead to Cow Island to meet the incoming boat and make reservations for the outgoing trip, leaving Isaac and Sternberg to solve the problems of moving the camp outfit. The fossils that remained in camp were loaded on packhorses, leaving as light a load as possible for the wagon. In order to make it from the deep valley to the prairie above, they had to build a makeshift road, bridging chasms with tree trunks and advancing step by step until late afternoon when they came to a ridge so steep it could not be climbed by going directly upward. They had to drive the horses at a sharp angle and inch up the slope. It was too much! When the horses and Isaac balked, the team and wagon began to roll sideways down the slope. After three complete revolutions, the horses landed on their feet and the wagon on its wheels as Sternberg stood in disbelief. Finally, however, they conquered the ridge and reached the open prairie. Even then, they had to go back on horseback and get the rest of their gear.

The trip to Cow Island took several days, all frightening and demanding on man and beast. The final push was pure drudgery and lasted from dawn one day until the next morning's early hours. Cope later wrote of this last day to his wife, but not wanting to worry her,

Steep terrain like this complicated fossil hauling.

The view from on high.

said simply: "We had a difficult task to get down to the Missouri
through the canyons and precipices. Had to let the wagon down with
ropes."[13]

While waiting for the steamer at Cow Island, Sternberg and Cope
made one last trip across the prairie toward some badlands, returning
in near darkness. Racing several miles toward the river, they tumbled
across ten feet (3 m) of cactus to a point where they could see the tiny
lights of the boat lying at anchor below them. Desperately they tried
to feel their way to a safe spot to descend to the water's edge. Cope used
a stick as a cane to prod the dark road ahead. Soon the stick met only
air, and stones Cope dropped into the darkness could be heard striking
far below. They had come to a precipice they could not traverse and
had to turn back and try again at a different location. Finally they
reached the river, but they knew they were not where they needed to
be; for the boat was nowhere in sight. They trudged four miles (6.4 km)
back up to the prairie and down into the next ravine. Sternberg, angry
and tired, longed to rest; but Cope pressed on, and at last they were
within shouting distance of the steamer. Cope wrote of the hazardous
descent, "At one point I made three attempts before I could get down
again to the high bank, and my horse came down some perilous places,
where he could only slide...."[14]

Cope shouted to the steamer's crew to send a boat to pick them up,
but the sergeant on duty feared it was an Indian ambush and refused
to come. Finally, Cope convinced the ship's crew to send a boat, but the
small craft overturned in the rapids. After the capsized men were
rescued and the boat righted, they tried again, this time reaching the

opposite bank. At last Cope and Sternberg were safely aboard the steamer where they ate a hearty meal of beans and raspberry jam on hardtack, and then fell asleep outdoors under a tarpaulin.

The next day, however, the captain refused to stop downstream to pick up the baggage and fossils, insisting that all cargo had to be loaded where the steamer was anchored—the ship would sail the next morning with or without Cope and company and their fossils.

Cope found an old scow anchored close by. After trying unsuccessfully to get permission to borrow it, he bought it for an exorbitant price (and sold it the next day). He and his men loaded seventeen hundred pounds (770 kg) of fossils onto the scow and then hitched their strongest horse to it with a long rope. Interested bystanders guided the boat and the horse steadily towed vessel and cargo upstream, finally reaching Cow Island. Cope wrote simply: "After sundry adventures in which we all got very wet and the horses rolled down the bank into a mud hole, we reached Cow Island."[15]

Cope, soaked to the skin and caked with mud, boarded the steamer and headed for Omaha with his precious cargo. Sternberg and Isaac broke camp, loaded the wagon and headed for Fort Benton, where they learned that Sitting Bull had crossed at Cow Island and killed the soldiers he found there.[16]

Sternberg bade Isaac good-bye at Fort Benton and never saw the man again. Ahead for Sternberg lay a brutal six-day stagecoach journey to catch the Union Pacific at Franklin, Idaho, then a long train ride back to Kansas for a few days with his family. After that would come another long train ride—to the East Coast to spend the winter with the Copes in Haddonfield, New Jersey, and Philadelphia.

Although the expedition netted few major discoveries, Cope described twenty-one species of dinosaurs from his collection of scattered teeth and broken bones. The trip had provided the first evidence of the rich variety of dinosaur fossils in the Judith River basin, and set the stage for the great dinosaur hunt that began the next year.

At a lecture in Philadelphia that winter, Cope displayed the skeleton of the Kansas mosasaur *Clidastes tortor* that Sternberg had found the preceding summer. Cope introduced Sternberg to the lecture audience as the man who found "this beautiful example of the fauna of the Cretaceous."[17] For Charles Sternberg, it was a moment of paramount glory. He was more conscious than ever of his lack of scientific education and of his inability to converse comfortably among men of science. At that moment, he felt Cope fully had repaid him for his long, dangerous, hard labor. Sternberg truly was on the road to success, and for all this Cope was responsible.

CHAPTER FIVE

A Secret Mission

The New Jersey winter of 1876-77 proved profitable for Charles Sternberg in many ways. Cope gave him a room in his workshop, and he "boarded out" for meals except Sunday dinners with the Copes. This was his first introduction to genteel life in the city. He spent many hours with Cope at museums and universities, studying, reading, and observing. Sternberg realized how much he had missed by not having a more complete college education with the connections and opportunities it would have presented. He could never learn as much about the science of paleontology as Cope and Marsh, but he vowed to find out as much as possible and be the best collector he could be.

He became aware of the strength of the relationship between Cope and his wife. Annie was a beautiful, gracious hostess and companion to her husband, and he took time to show his affection for her. In the Sternberg home, Levi had never seemed to have time to spend with Margaret and she, for her part, was so busy with her large family and the responsibilities thrust upon her by her minister-teacher-husband that there was little time for kissing and embracing.

Charles watched the Copes in their special moments when the rest of the world seemed to vanish, and they had time only for each other. Cope had vicious headaches, and Charles saw Annie tenderly massage away his tensions and pain, helping him to relax and gain new vigor while he rested in her embrace. Little Julia Cope was a delightful, well-trained, and loving child who brightened the home with her sparkling exuberance.[1]

Charles thought of his own childhood, and the contrast made him realize he was missing something important in life. Maybe he needed a woman—a wife and companion. The thought had never really occurred to him before. But the more he considered it, the better the idea seemed! But where would he find a woman? How could any woman share the life he led? What did he have to offer?

Such thoughts returned again and again on the long train ride back to Kansas in the spring of 1877. Sternberg thought about the few girls he had known around Ellsworth. None of them seemed worth remembering. There had been a few girls in Manhattan in his classes, but he couldn't even recall their names, much less where they came from.

The train stopped in Lawrence. He remembered this was the home town of the chaplain, Rev. Charles Reynolds, whom he had known at Fort Riley. Then he thought of Anna who had come with her father to visit him in the fort hospital. Anna was a pretty girl! The train stopped in Fort Riley, and he fidgeted a bit, then sank back in his seat. Now wasn't the time. He had to get home and think about it first. The train chugged west toward Ellsworth. Ten years had come and gone since he had first come to Kansas as a mere stripling, eager for the excitement of the frontier. He'd be twenty-seven years old this June. He hadn't accumulated anything and didn't have much to offer a wife, except his ambition to be the best fossil hunter in the West. A woman might find it hard to live on what his fossils brought in. He gave a short laugh, gathered his luggage, and stepped off the train in Ellsworth.

Charles spent a few days with his parents, escaping whenever he could to tramp through the rock-strewn countryside. He found more sassafras leaves, then stumbled onto a skull of a creature he couldn't identify. He took up the specimen, packaged it carefully and shipped it to Cope with this letter:

> Wilson, Ks
> Apr. 3d 77
>
> Prof E D Cope
>
> Dear Sir
> We arrived here last night. I found a skull of some extinct animal which I purchased. I send it to you today per express. I did not have the money to pay for it so please send it [the money] to Mr. Jacob Fowler. He would like to know what it is & the price is $2. Please send it at your earliest convenience...
>
> Yours truly   [signed] Charles Sternberg[2]

Not waiting for an answer, Sternberg prepared to return to the chalk beds in western Kansas. He hired helpers to accompany him and headed for his old base at Buffalo Park. It was late in April. The next weeks were uneventful until one day, south of Hackberry Creek, he found large shells of the Cretaceous bilvalve mollusks *Haploscapha*, or *Inoceramus*—some almost a foot (30 cm) in diameter—as well as many fish, and aquatic reptiles. He was encouraged, but then his plans changed.

Early in July, Cope sent word and money for Sternberg to go north to the forks of the Loup River in central Nebraska to hunt vertebrate fossils in the beds of the Upper Miocene called the Loup Fork Group by Hayden. But while making preparations in Buffalo Park, Charles visited with a venerable hunter who had just brought in a load of buffalo hides. The hunter described a mastodon's skull he'd seen

A large Haploscapha *shell, found before 1900 in Kansas chalk by Charles H. Sternberg, as displayed in the Sternberg Museum, Hays, Kansas.*

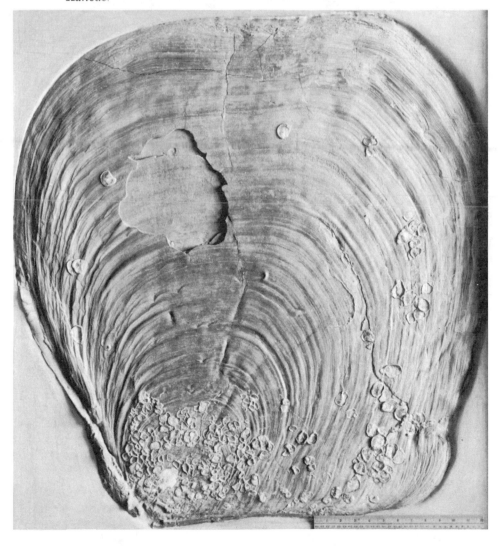

protruding from solid rock near his cabin on the middle branch of Sappa Creek in northwestern Kansas. Sternberg decided to make a short detour on his proposed route north and have a look. He discovered enough of interest there that he did not cross into Nebraska all summer. It was an area he profitably returned to examine at greater length a few years later.

At sundown one evening, Charles and his men found a great land turtle thirty inches (76 cm) long, eight inches (20 cm) across, and fifteen inches (38 cm) high.[3] It was the first notable discovery he had made that summer. But soon thereafter he found specimens of the inferior tusked mastodon *Gomphotherium*, originally named *Trilophodon campester* by Cope. Then he found quantities of rhinoceros bones from a species that Cope later named *Aphelops megalodu*s. The excitement of this discovery was enhanced by the fact that the bones, embedded in beds of soft limestone and sandstone, were easier than usual to remove.[4]

In August Sternberg received orders from Cope to go at once to a new field in the desert of south central Oregon, but the trip and destination were to be kept a secret. As in a treasure hunt, the directions said: "Go to Fort Klamath, Oregon, and from there to Silver Lake, to a man by the name of Duncan, the postmaster. He will guide you to the fossil bed in the heart of the sagebrush desert. You will likely find human implements mingled with extinct animals. You are to go secretly; tell no one where you are going. Have your mail sent by a circuitous route, so you cannot be traced."[5]

Charles closed down his operations in Decatur County, Kansas, and rode horseback the seventy-five miles (120 km) to Buffalo Park Station where he took the train to Ellsworth. He thought briefly about the chaplain's daughter, but he had no time now to visit Fort Riley. He did have a few hours with his parents, but said nothing to them about wanting to find Anna Reynolds. In fact, he wasn't sure he was ready for romance. It could wait. He packed his bag for an indeterminate stay in an unknown area, with little idea of what he might need or would encounter.[6]

Only twenty-four hours after leaving Buffalo Park Station for Ellsworth, he returned, ready for his new assignment. He loaded his tools, blankets, and baggage, and boarded the westbound train only to find his friend Samuel W. Williston, a member of Marsh's crew, also on board. Each tried to pump the other for information, but both were well coached in the art of deception and secrecy. Their friendly banter cooled, and when the train pulled out, silence hung heavy between them. Their long friendship, begun in Manhattan ten years earlier, finally resurfaced, however; and they spent the night amicably in Denver. Williston headed south without disclosing his destination;

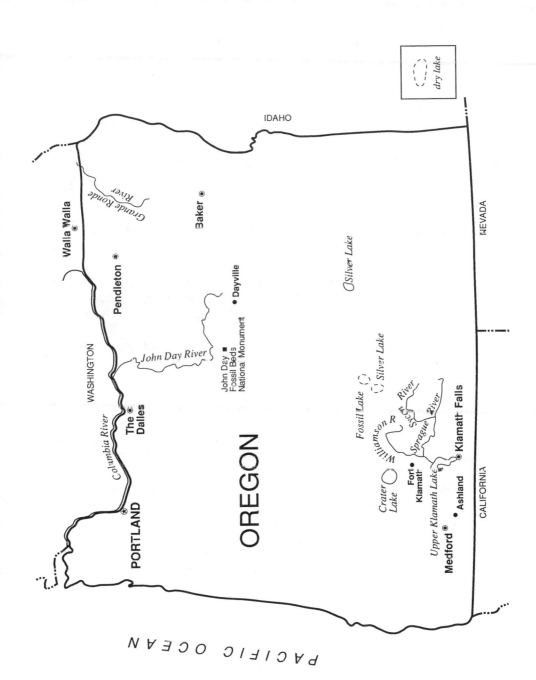

and Sternberg, also quiet about his plans, continued west, smug in the realization that he had kept Cope's confidence intact.[7]

Charles left the train at Ashland, Oregon, and completed the journey east to Fort Klamath on a buckboard pulled by a team of ponies. At Fort Klamath he hired George Loosely as an assistant and bought two saddle ponies and a packhorse, a government tent, and rations. He lingered long enough to get his flour baked into bread by the post baker, then started for Silver Lake, although no one at the post could give any directions.

Sternberg and Loosely tried to follow the map Cope had sent, but were confused. They had to cross a government bridge on the Williamson River, then go to where the road ended in an Indian village on the bank of the Sprague River. At the Williamson River, a Snake Indian stopped them and demanded toll at the bridge. Believing that as American citizens they had paid taxes to build the bridge and therefore did not owe the toll, they refused and rode across despite the Indian's threats. That evening they were near the Sprague River and went into camp near a large Indian town. Loosely learned that a chief was dying, and his curiosity got the best of him: he wanted to witness the death ceremonies. Before taking leave of Sternberg, who wanted only to get to bed and rest, he helped stow the bread and coffee between the mattress and blankets and hid the bacon at the bottom of the mess box with tin dishes piled on top. Assuming their food was safely hidden, Loosely left and Sternberg crawled into bed.

When Loosely returned at about 3 a.m., he was exhausted. The Indians had danced in a circle around their chief in a closed room all night and forced Loosely to join in. Sternberg roused when Loosely came in, but soon both men were sound asleep. A few hours later Loosely put water on to boil for coffee, and learned that the cache had been broken into: the bacon was gone; the dishes rearranged as though undisturbed. The men had to settle for bread and coffee, and left a bit disgruntled and sheepish that they had slept so soundly they were unable to protect their food.

Early that morning they saw the first white man they had seen since leaving Fort Klamath. They asked him for directions to Silver Lake and were told to go north on the trail to a sheep camp in Sycan Valley where they would receive further directions. At last Sternberg felt he was on the right trail, as outlined in Cope's letter. Now they needed to find Silver Lake and a man named Duncan.

They rode all day and as evening approached entered a forest only to find a fork in the trail. While they pondered which direction to take, a boy appeared driving two pack ponies. Sternberg asked him where he was going. "To Sycan Valley, to a sheep ranch," he answered. Cope's secret code had worked. Gratefully Sternberg and Loosely followed the boy. One more day of travel brought them within earshot of the bleating of sheep, and shortly thereafter to the sheepherder's camp, where they spent the night. The next day they found the postmaster, Mr. Duncan, and his family relaxing in his home on the shore of Silver Lake.[8] Duncan loaded his wagon with Sternberg's supplies and took him and Loosely through the sagebrush to a small alkaline lake in the heart of the desert, some fifty miles (80 km) west of Silver Lake. That night they stopped at the home of a hermit rancher named Lee Button and picketed the ponies on a grassy flat. Duncan dug up a tin can that held a key to the house. Button was not at home, but by prearrangement with Duncan, his house was available, as was the custom of that time. The pantry was well stocked and sagebrush lay stacked by the fireplace. Pungent fumes from the hot fire soon filled the room; and after a hot meal the weary travelers went to bed.

Wild geese roamed around the small lake nearby, some straying quite close to the house. Rather than waste ammunition on an easy shot, Sternberg chose to trap food for the next day's meal. By morning he was rewarded, finding a brant—a variety of wild goose—in his trap.

The trail ended at the Button ranch, so the rest of the journey was uncharted. Duncan, however, knew the area and skillfully guided the party into and around small hills, using the mountains in the west to help confirm directions. That evening they came to another small alkaline lake. "There," shouted Duncan as he pointed with his whip to the lakeshore, "There is the bone yard."[9]

The clay bed of the lake, which Sternberg dubbed Fossil Lake, was strewn with numerous bones and teeth of reptiles, birds, and mammals—truly a bone yard! Bones were so scattered that no two seemed to belong together. Many skulls had been crushed. Abundant arrowheads and spear points of polished obsidian or volcanic glass lay among the fossils. Quickly Sternberg filled a cigar box with fossil teeth, arrowheads, and spear points, addressed the box to Cope, and sent them off with postmaster Duncan.

For several weeks Sternberg searched for fossils in the lake bed and the surrounding hills. He found a fossil swan, flamingo, heron, six genera of fish, and fifteen species of fossil mammals, including llamas, horses, elephant, dog, otter, beaver, and great sloth.

Finally Sternberg felt he had explored all the Fossil Lake area and wanted to move on to a new region. He spied the top of a dead spruce tree rising above a sand dune and climbed a hill for a better look. He realized he was looking down into a valley created by the force of winds blowing relentlessly across the sand. He made his way down to the valley and found the remains of an Indian village. Near where there had once been lodges, he found great piles of bleached animal bones—neither petrified nor fossilized. He found a spring of fresh water and

This slab of fossils, recovered from a quarry in Weld County, Colorado, illustrates the unusual collection of bones sometimes found in one pit.

a number of arrowheads and spear points, which he quickly collected. Then he noticed that daylight was almost gone. He headed his pony toward the Button ranch and had to depend on the horse's homing sense to find the way as night closed in.

He began to shout, hoping Loosely would answer. After several futile attempts he finally heard a response. Loosely rode out and guided him back to the Button ranch.

A few days later, when Sternberg and Loosely were hunting, they found some mud springs—great circular areas that when wet held yellowish muck, the consistency of mortar. Most of the time, however, the areas appeared dry and hard. All of a sudden the packhorse, heavily laden with the day's take of fossils, went down in the thick mud. His thrashing only made his predicament worse. In desperation, Sternberg cut the ropes that bound the bags of fossils and camp gear and, with Loosely's help, pulled the precious cargo to safety. They then turned their attention to the struggling animal and, by tying a rope around his neck and tugging on it together, pulled him to safety. They led him to a tiny stream where they washed him as best they could, then headed for the Button ranch. They fed and bedded the horses down in the barn, helped themselves to good food from Button's panty, and finally lay down on Button's blankets, totally exhausted. Someone knocked at the door, then came in. The stranger said he needed shelter for his team, food, and a place to sleep. Could he stay?

"Why certainly," Sternberg answered. "I don't own the ranch, but we have just put our horses in the barn, where there is plenty of hay and oats, and there is plenty of food here. George [Loosely] will show you the way to the barn and help you unhitch, and I will have supper ready when you return."

"Thanks."

The stranger put up his horses and enjoyed the meal; but during conversation later in the evening, Sternberg had a suspicious thought. "Do you know Lee Button?" he asked.

"Yes, I've seen him."

"That's your name, isn't it?"

"Yes."

Sternberg was embarrassed and apologized for the freedom he had taken with Button's property, but was assured that that was just what Button wanted to have happen. The incident developed into a lasting friendship and home away from home for Sternberg.[10]

Cope sent word for Sternberg to remain in the West that winter and to be ready for further assignment in the spring. He established a camp on Pine Creek, Washington, and explored swamps and struggled against water—quite a change from the desert that had challenged him all summer. He and Loosely dug a shaft down about twelve feet (3.6 m) to a bed of gravel where they found fossils, but every night the shaft filled with water, and the process had to be repeated each morning. Sternberg said he was never dry all winter.

Taking some time off from fossil hunting, he went to Fort Walla Walla, Washington, where Dr. George Sternberg was post surgeon. Previously, George had been stationed at Fort Barrancas, Florida, where he contracted yellow fever with complications. Near death for weeks, he finally recovered sufficiently to go on leave and spent the better part of year with his wife in Europe, regaining his strength. Upon his return to duty, he was sent to the West Coast and eventually assigned to Fort Walla Walla.

Charles had not seen his brother since 1870 when George had been transferred from Kansas to New York. Now, with the return of good health, George had found time and opportunity to do a little fossil

57

hunting himself, so he and Charles had much to talk about during the cold, dark days of winter. One escapade they discussed involved a discovery made by George's friend, a civil engineer who also enjoyed fossil hunting as a hobby.

When the friend told him about some amazing specimens he had found not far from the fort, George and his wife had organized and outfitted a party and headed northwest to the banks of the Snake River near where it empties into the Columbia. As they made camp, they heard tom-toms and then saw several boats full of Indians coming across the river toward their campsite. A chief and an interpreter disembarked and approached. They wanted a medicine man, and finally George learned through their interpreter that the chief had a sick daughter who had been coughing "for two snows." Could the white medicine man help her? Guessing that such a persistent cough indicated tuberculosis, George knew there was little he could do for the young patient except give her temporary relief. This he did, but the interpreter asked for more—coffee, sugar, and other foodstuffs.[11] Eventually the Indians were appeased, but the doctor was apprehensive. He was in Indian country and knew he probably would have more encounters.

The party moved on the next day toward Washtuckna Lake where George's friend had said he had found fossils. While Indians hovered not far from camp, the Sternbergs set about exploring the lake area and were overwhelmed by their discoveries, including whole skeletons of horses, elk, and deer. George theorized that the animals, thousands of years earlier, perhaps had gone to the nearby lake to drink and sunk helplessly into the morass of sand along the shore where they died from suffocation and starvation.[12]

George sent some of his fossil treasures to Cope and other friends interested in paleontology. Part of his collection he saved and showed to Charles, then kept for the rest of his life.

Although George's party wanted to continue the hunt, the constant presence of Indians prompted them to break camp and head back to Fort Walla Walla as quickly as possible. The restless Indians were Nez Perces, who had been friendly and helpful to early settlers, but had later become hostile and militant under their leader, known as Joseph to white men and as Halla Kalla Keen—Eagle's Wing—to his own people.[13]

George's experiences provided long hours of exciting conversation for the Sternberg brothers that winter, and Charles was delighted to see George's fossil collection. He knew that time and distance would never really separate them and was grateful for the chance to spend time in his brother's home. He found Mrs. Sternberg totally charming and realized how important home life was to this brother. Again the thought of marriage crossed his mind, but the time was not yet right. He ventured a cautious question or two about the chaplain back in Fort Riley. Did George know whether he was still there? Did he know what had become of the chaplain's daughter who had been so nice to Charles at the hospital? Nothing came of the conversation. Charles simply stored the answers in his mind for action when he could get back to Kansas.

Charles left Fort Walla Walla the third week in April 1978, traveling with a team and wagon and accompanied by his assistants Joe Huff and "Jake" Wortman, a young man he'd met through George and with whom he had camped part of the winter after Loosely took off to pursue other interests. Unknown to Sternberg at the time, Wortman was a scientist of renown and would continue to figure in his life for years to come.

The group headed southwest and after fifteen days reached Dayville on the South Fork of the John Day River near the intersection of modern-day US 26 and Oregon 19. Here, at the foot of a canyon several thousand feet deep, the mountains swung away from the river in a horseshoe bend and closed in again a few miles below, forming an amphitheater three miles (4.8 km) wide and thirteen miles (20.8 km) long.[14]

Sternberg gazed in amazement at the vastness and beauty around him. The horizon glowed with the green, yellow and orange of the clays and volcanic ash beds of the Miocene. The geysers and fumaroles

Gorge of the John Day River of Oregon, which yielded many of its fossils to Charles H. Sternberg. —Oregon Historical Society, Negative #84826

emitted steam and held him spellbound. The rugged cliffs appeared almost unsurmountable, but Charles never turned back! With a collecting bag over his shoulder and a sturdy pick in hand, he and his assistants worked their way up the cliffs, scanning for a point of a tooth, an end of a bone, or a concretion that would reveal the hidden fossils.

They found many fossil leaves—oak, maple, and others—as well as fish vertebrae and pumice, a volcanic rock so light it floated in the streams. The terrain was exciting, the river swift and dangerous. An old hermit pointed out the volcanic cliffs to Sternberg, assuring him of a wealth of fossils nearby. Sternberg picked up another helper, Bill Day, who had worked for Marsh and was familiar with the terrain. Sternberg was somewhat surprised to discover that the area had been explored before. His secret destination, as outlined by Cope, was not virgin territory after all!

Making their way to the river's edge, they were dismayed to discover that the real fossil beds were on the other side. Sternberg found an old boat caught in a pile of driftwood and dug it out with bare hands, only to find that its hull was a veritable sieve. Using sticky clay, he caulked the seams and holes and managed to get to the other side of the river before the caulking softened and nearly sank the boat. He secured it for the return trip and climbed the bank to search for fossils. He found many fine specimens, including the skull of an *Oreodon*, an extinct porcine animal common in the area during late Oligocene times.[15]

After recaulking his boat, Charles returned to camp and ordered Joe Huff, who owned the horses, to prepare to cross the river. Huff feared for the safety of his animals and refused to make the trip. Sternberg paid him his wages, and Huff left, riding one horse and leading the others. Disgruntled by the inconvenience, Sternberg delayed his hunt until he completed arrangements with Day, who owned a herd of ponies and had promised to supply as many as Sternberg needed.[16]

Sternberg discovered that other fossil hunters had been through the area, as he found only scattered bones and fragments. The cliffs were so steep that when they found a fossil they first had to cut an area large enough to stand upon before they could carve out the specimen. Every day was a new adventure, its outcome uncertain.

One day Sternberg found the cylindrical foot bone of a large camel. Concentrating on a search for more camel bones in the older fossil beds, he discovered a skeleton of an unnamed species in bold relief on the face of a slope. Cope would call the species *Paratylopus sternbergi*.[17] Wortman later found a skull of this species, and both specimens were placed on exhibit in the American Museum. Sternberg was exultant.

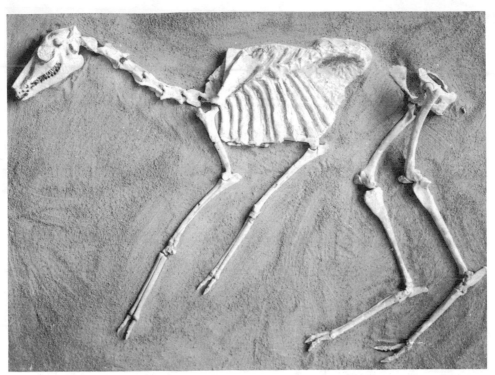

Small fossil camel.

His summer was going well, despite complications and reversals. It was only June, and he had taken out many fossils, including: a skull of the rhinoceros *Diceratherium namum marsh*; a skull of a peccary; an *Oreodon*; and a dog about the size of a coyote that Cope named *Enhydrocyon stenocephalus*. Altogether in the John Day basin that year Sternberg found hundreds of specimens, many of which were stored in trays at the American Museum. These included about fifty species of extinct mammals.

One day in July Sternberg started alone for Dayville on horseback, leading a pack pony. He planned to spend the night with a friend and take back a load of provisions. Just as he reached the summit of a mountain above Dayville, he looked down into the narrow John Day River valley and realized something was wrong. There was a village, but no life, no smoke rising from the chimneys. Wheat in the fields was ripe, but no one was working. He followed a trail down the mountain, shouting occasionally, hoping someone would reply. When he finally reached the river, he was delighted to see his friend, a Mr. Mascall, who had a home and a second cottage, which Sternberg had sometimes used, across the river in the valley. Mascall worked rapidly to load Sternberg's gear and saddle, but kept silent until they were on the other side of the river; he then explained that everyone had gone to Spanish Gulch, a mining town on top of the mountains about ten miles (16 km) southwest, because Indians were on a rampage. Only Mascall and an old man who kept the mail station had remained behind. That evening Bill Day arrived, having heard the news at an Indian camp and wanting to help Sternberg escape.

About three hundred Bannock Indians had started north from the Malheur Agency, several hundred miles south, to join the Umatilla at Fox Prairie. Sternberg insisted on returning to his base camp to find Wortman and made the dangerous trip alone. He found Wortman hard at work removing a fossil he had just uncovered. He and Sternberg hastily took down the tent and hid it and all the fossils stored in it under heaps of brush. Then they rode rapidly back to the river where Mascall and Day met them. Here again, Sternberg had a cache of fossils he couldn't bear to leave. They used lumber from a nearby mill to make small boxes, working through the night with one man keeping watch for signs of approaching Indians. By daylight the fossils were packed into the boxes, carried down to the river's edge and hidden under a great grapevine. They gathered leaves and twigs to conceal the spot further and to cover their trail.

Mascall refused to leave his home; the fossil hunters reluctantly bid him good-bye, realizing they could do no work in the area for a while. Sternberg decided to go to The Dalles, more than a hundred miles (160 km) northwest, to find out what had become of a collection of material from Fossil Lake that he had ordered shipped to Cope in Philadelphia. The shipment had never arrived, and attempts to trace it had been futile. He began the long trip on horseback, alone, then joined a party of men, women and children also fleeing Indians. For a few days he rode with a party taking several hundred horses to a safer area, but the dust became so intense that he decided he would be better off alone. Striking out again by himself, he reached The Dalles where he learned from the Oregon Steam Navigation Company, which was supposed to have shipped his fossils, that they were still there, covered up in a warehouse and forgotten. He waited until he was confident the shipment was on its way, then started his long trip back. A Leander Davis, who had hunted fossils for Marsh some years earlier, volunteered to accompany Sternberg; and together they packed their horses and began the journey back southeast to find Mascall and Day.

They continued east, crossing the South Fork of the John Day River, and soon found signs that Indians had passed through. A wide trail cut deep into the soil by horses' hoofs and by Indians on foot left no doubt in their minds of the dangers around them. A short time later they saw horsemen approaching; General Howard and his staff were seeking a place to rest and find food. Only a short distance ahead the group found a smokehouse, unoccupied, but with evidence that the former occupants had left suddenly just before sitting down to a meal. There was cold coffee in the cups and cold food on the table. Davis and Sternberg left the soldiers there and continued on their journey, finding house after house where the Indians had killed settlers, mutilated animals and destroyed possessions. When they reached Mascall's place, they retrieved the fossils, packed the horses again, and headed for Haystack Valley where they established a camp.

The uprising, known as the Bannock War, ended dramatically in late summer of 1878, restoring peace to the valley. Cope ordered Sternberg to remain in Oregon for another winter. The next spring he again explored the John Day basin, sending his fossils to Cope. One fossil was acquired in memorable fashion there that spring of 1879 when Leander Davis recovered the skull of the saber-toothed tiger *Pogonodon platycopis*. The skull lay at the top of a pinnacle, estimated by Sternberg at thirty to forty feet (9 to 12 m) high. Tapered like a church spire, the pinnacle was too steep to climb and too fragile to support a ladder, even if they'd had one long enough. After weeks of painstaking work on the part of Sternberg, Davis, and Wortman, Davis threw a rope around the pinnacle and inched his way, hand over hand, up to where he could reach the skull and carry it back down the

Fossilized log in the John Day River country of Oregon. —Oregon Historical Society, Arthur M. Prentiss photo, Negative #84827

treacherous descent. The skull went to Cope, who wrote of the dramatic achievement.[18]

It was time for Sternberg to go home. He had spent two frightening and eventful years; he had explored Oregon from the southwest to the northeast, throughout the center two-thirds of the territory. He was doing what he most wanted to do with his life. But he now knew home and family were important too. He wanted to return to Kansas, visit his parents in Ellsworth—and maybe even go to Fort Riley to see if Chaplain Reynolds and his daughter were still there. As he rode the train home that summer of 1879, he philosophized about what he had been doing:

> What is it that urges a man to risk his life in these precipitous fossil beds? I can answer only for myself, but with me there were two great motives, the desire to add to human knowledge, which has been the great motive of my life, and the hunting instinct, which is deeply planted in my heart. Not the desire to destroy life, but to see it. The man whose love for wild animals is most deeply developed is not he who ruthlessly takes their lives, but he who follows them with the camera, studies them with loving sympathy, and pictures them in their various haunts. It is thus that I love creatures of other ages, and that I want to become acquainted with them in their natural environment. They are never dead to me; my imagination breathes life into "the valley of dry bones," and not only do the living forms of the animals stand before me, but the countries which they inhabited rise for me through the mists of the ages.[19]

CHAPTER SIX

Christmas Blues

C harles returned to Ellsworth in the fall of 1879 and located Anna Reynolds. Still single, she was living with her mother in the family home in Lawrence. Her father still served as post chaplain for Fort Riley. Charles courted Anna gingerly. He felt awkward, as he had few worldly goods to offer her. Anna's response was better than he had dared hope. Slowly their friendship blossomed into love, and they married July 7, 1880, in the chapel at Fort Riley, with Anna's father, the Reverend Charles Reynolds, officiating.

Levi Sternberg approved of his son's bride and in-laws, hoping marriage would bring about a change in his lifestyle and ambitions. He welcomed Anna into the family and prayed for their future.

The Reynolds family, for its part, accepted Charles and invited him to bring his bride to their home since he had no place of his own, nor any prospect of establishing one in the near future. Anna lived with her mother while Charles was off hunting fossils. Someday, he hoped, he would provide her with a home of her own. For the present, however, he joined Anna in the Reynolds home for a few months each winter and was relieved to know she was not alone while he was gone the greater part of the year.

Residents of Lawrence no longer had to fear Indian attacks, but the memory of local violence remained fresh in the minds of all residents. Anna told Charles of her family's experience when Confederate guerrilla William Quantrill led a raid there in 1863, burning much of the little city and massacring 150 unarmed men and boys. According to the story later handed down to her children and retold by her son one hundred years after the raid, the guerrillas came to the house, where Anna's mother gave them freshly baked bread and milk. Only moments earlier, she had taken in a frightened neighbor disguised as a woman. The disguise fooled the raiders, despite the neighbor's

No photo exists of Anna as a bride. In this earliest surviving portrait from about 1912, she was already fifty-nine years old.

Charles H. and Anna lived in her parents' home in Lawrence for the first years of their marriage.

crudely shaven face and a slip-up by one of the children, who screamed at a menacing raider, "Don't touch him. He's our old auntie!" The Reynolds family and house were left undamaged.[1]

For two years Charles was content to hunt fossils in the Niobrara chalk of Kansas, returning to Lawrence and Anna when winter closed down his field work. He set up a workshop in an abandoned building in Lawrence where he worked long hours cleaning and preparing the specimens he had shipped in from the field.

In May 1881, Anna give birth to a son, Charles Reynolds. The young couple's joy was not to last, however: their son was fated to die in September 1882.

In the summer of 1882, the Museum of Comparative Zoology, Harvard, hired Charles to hunt for fossils in the Upper Miocene "Loup Fork Group" in north-central Kansas under the direction of Dr. Ferdinand Hayden. The group included an area extending from south of the Platte River in southwest Nebraska into the Solomon River valley in northern Kansas.

Before going West with Cope in 1877, Sternberg had found numerous fossils—including a mastodon and bones of three-toed horses—in the Ogallala formation of the same area. Now, as he worked the area again, he studied the terrain carefully and determined the outline of a promising region near Prairie Dog Creek, close to the little town of Long Island, Kansas. Here lay vast quantities of assorted fossil bones—a veritable gold mine!

As he looked for a place to pitch his tent, Charles found a large exposure of hard gray sandstone. Just above this rocky ledge lay a sand-filled quarry. Rhinoceros bones protruded from the sand, and a ditch at the bottom of the quarry exposed rhino toe bones, broken skulls, and teeth. He was overwhelmed. Never in his seventeen years of fossil hunting had he found such a great deposit in one place! He took possession "in the name of Science, of the largest bone bed in Kansas"[2] and named it Sternberg's Quarry. It did not occur to him that anyone else would take any other view of the claim.

His triumph was short-lived, however. An old man, plowing corn along the eastern edge of the area, saw Charles working with his pick. After watching him uncover the skull of a rhino, the farmer shouted in anger, "What are you doing?"

"Digging up antediluvian relics."

"Well, get out of there!"

"All right," Charles answered, but he kept on working.[3]

The old man left, and Charles thought the incident was over. But when he went into town for food he learned that the farmer had gone to a justice of the peace to request a warrant for Sternberg's arrest; his offense: collecting the old bones. Sternberg later learned that the old man had gone all over that part of the country trying to get the warrant. Fortunately, the farmer was not successful.

One day as Sternberg rode past a farmhouse, he paused in answer to a shout and wave from an old gentleman and his wife. The two were sitting in a makeshift cart concocted from a dry goods box with two wagon wheels attached. A team of ponies with a rope harness and lines pulled the little cart. The old man, sitting tall and straight, demanded to know what Sternberg was doing. "I'm looking for rhinoceros bones in the loose sand of the hills here," he answered.

"Well," the old man continued, "I am interested in these old bones myself. I don't claim to be a scholar. In fact, I am quite illiterate, but I think when this earth was in a molten state, these old hippopotamuses wallowed around in the mud and got congealed in the rocks."[4]

Another afternoon Sternberg found a complete skeleton and shell, four feet (1.2 m) in diameter, of the turtle *Testudo orthopygia*, originally described by Cope. Unfortunately someone had chopped them to small pieces. A little farther on he found another skeleton, then another. All had been damaged the same way. He returned to camp and lashed out at his assistant: "Some infernal vandal has been up this ravine and dug up with a mattock three of the finest turtles I ever saw."

A stranger who just happened to be lounging on a box at the camp jumped to his feet, hung his head, and then nodded. "I was out here digging roots to build a fire with and ran across them. I didn't know they had any value, and I wanted to see what was inside of them and dug into them."[5] Sternberg was dumbfounded. Everyone was silent. Sternberg at last saw the humor in this episode and broke the tension with a hearty laugh.

A great land turtle found in Loup Fork Miocene, Phillips County, Kansas.

After Sternberg's first summer in his own quarry, he returned to Lawrence in the fall of 1882. He left again in late December, this time to collect fossils in the Permian beds of Texas for the Museum of Comparative Zoology, Harvard. It was an exciting, new assignment, and Charles set forth determined to make a good showing and establish a new outlet for his discoveries. He was working now as a free-lancer on contract, free to find his own markets and employers. He reported first to Harvard, then left Cambridge in the middle of December and arrived in Dallas a week later.[6]

At Harvard, he had been given the name and address of someone to contact in Dallas, but when he arrived there, he found no such person or address. He had no idea where the Permian beds were. Hayden had told him to follow the Red River until he found red beds, but Sternberg had seen red mud as soon as he entered the state and knew it was impossible to cover the entire area that winter looking for the Permian rocks.[7]

Finally, he went to the post office where the postmaster referred him to a real professor, W. A. Cummins, who had worked in the Permian beds with Cope the year before. Sternberg didn't find Cummins, but secured accurate directions from Cummins's wife. She had been with her husband the previous year and told Sternberg to go first to Gordon, the closest station on the railroad, then take an overland route north and a little west to Seymour. Confident he could find the right location and secure sufficient help, Charles had no qualms as he boarded the train for Gordon.

It was Christmas Eve, and he fought off a wave of loneliness and sorrow as the train neared the tiny town of Gordon. His firstborn son, only seventeen months old, had died suddenly in September. Anna was with her family in Kansas, but he longed to be with her. They had spent only two Christmases together. While he brooded over his situation, the train stopped and he detrained, the only passenger to do so. Almost immediately he was welcomed by about twenty boisterous cowboys, full of holiday cheer.

"Where are you from?"

"Boston."

"Where do you want to go?"

"To the best hotel in town."

"All right, we'll take you there."

They formed double line with Sternberg in the middle. Two men placed their guns over his shoulders, and the two behind him crossed these guns with their own. Then came the command: "Fire at will!" The shooting began and did not stop until the procession reached the hotel. When Sternberg was safely inside the hotel in custody of the chambermaid, the half-drunken cowboys continued their rowdy march and soon were out of sight.[8]

Sternberg hired George Hammon, the son of the hotel keeper, to load the baggage in a wagon and drive him to Seymour. The trip took eight days, crossing relatively flat plains punctuated with streams and small lakes, but chiefly red soil covered with mesquite and sagebrush. The land slopes to the north where the Red River divides Texas and Oklahoma. Sternberg found the Brazos River bed 175 feet (52.5 m) higher than the Wichita and in between, as they approached the breaks of the Wichita, they saw miniature badlands with rounded knobs, deep canyons, bluffs, and ravines. Deep gulches cut through the landscape, and mounds capped with white gypsum rose from the red clay.

The water here was alkaline, so men and animals depended on rainwater that pooled in the red clay. Wind and water had scoured dish-shaped basins in the clay, making it a good bed for natural water storage. Sternberg found that he could clarify the water by adding pulp of cactus leaves and boiling the mixture. He soon came to accept an old cattleman's theory: "Wherever and whatever a cow can drink, so can I."[9]

For six weeks, Charles searched fruitlessly for fossils. He saw thousands of wild turkeys, which became his most reliable source of food. In mid February he went below the forks of the Big Wichita and found some strata containing small, silica-cemented nodules. These multicolored nodules held fragments of the amphibian *Eryops* and the long-spined reptile *Dimetrodon*. Charles collected seventy-five pounds (34 kg) of *Dimetrodon* bones preserved in iron ore concretions.[10]

All winter he worked the desolate beds, walking over thousands of acres of naked rock. Storms came up suddenly in this open country. One January day while out in the field, Sternberg had seen cattle thundering across the prairie, heading for timber. They had scented a "norther." Charles hurried back to his camp, where he cut poles, stretched his wagon sheet across them, and secured it to the ground on both sides. He now had a makeshift shelter, open at both ends. Charles dragged some fallen wood under cover and built a fire near the entrance to his tent. Barely finishing his task, he watched the storm unleash, first with relentless wind, then darkness, snow, and sleet. For three days and nights the storm roared; Charles huddled near the dying fire, exhausted, discouraged and miserable. Hammon had tired of his job, taken his horses, and driven off, leaving Sternberg stranded thirty miles (48 km) from town! Fortunately, he had met a good-natured Irishman, Pat Whelan, who had horses and agreed to help. Sternberg had sent him to town just before the storm broke and had not heard from him since.

Finally the storm spent itself and a great quiet engulfed Charles's makeshift camp. He had kept warm and dry, and now, for the first time since the fury began, he stepped outside to survey the damage. Whelan finally returned; he had lost his horses in the storm and had to find others. He had agreed to stay with Charles only until time for spring farm work and would soon leave.

Charles also battled emotional storms that winter—discourage-ment, homesickness, and loss of helpers. He had found few notewor-thy fossils. Finally, by the middle of March, the weather moderated. He remembered he had a letter from the secretary of war, obtained through the efforts of Professor Alexander Agassiz and signed by Robert T. Lincoln, addressed to commanders of western posts, re-questing them to assist the bearer in every way possible.[11]

Doctor Hayden at Harvard had told him to search for fossils in the red beds near the Red River. Despite all efforts, he still had not found these beds. Now he was alone and he needed help. He decided to go to Fort Sill, north of the Red River in Oklahoma Territory where he hoped he could secure horses and men who would see him through the last weeks of his assignment. Without Whelan and his horses, Charles was on foot! He went to a local livery stable and rented a pony; he also contacted a new helper who agreed to stay at the base camp until Sternberg returned from Fort Sill, about sixty miles (96 km) away.

At the livery stable, Charles was assured that the trip could be easily completed in one day. But this was not the case. At nightfall, he

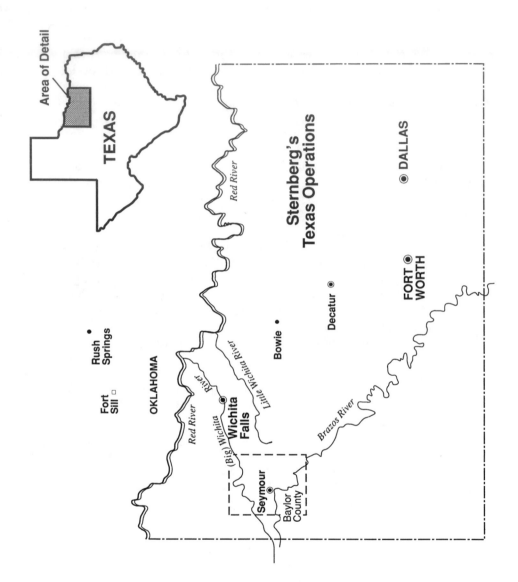

Area of Detail

TEXAS

Sternberg's
Texas Operations

Red River

● DALLAS

Rush
Springs ●

Fort
Sill □

OKLAHOMA

FORT ◉
WORTH

Decatur ◉

Bowie ●

Little Wichita River

Red River

(Big) Wichita ◉ Wichita
Falls

Seymour ◉

Baylor
County

Brazos River

was only halfway to his destination. He found shelter with a small encampment along the trail, pulled out his letter from the secretary of war, and received good food and a place to sleep. The next day he crossed the Red River, but again had to stop for the night, this time with friendly Indians. An Indian boy walked a few miles with him early the next morning, smoking a cigarette made from a dry tobacco leaf Sternberg gave him. When he finished smoking, he dropped the burning leaf to the ground, watched it catch grass on fire, then stomped it out. Charles thought little about it until later when Maj. Guy Henry at Fort Sill asked, "Did you leave the crossing at Cache Creek about sunrise yesterday?" Sternberg answered yes and the major replied that probably ten or fifteen minutes after his departure from the Indian camp, the Comanche chief had received notice by smoke signal that one man was coming over the trail toward the fort.[11]

Again the letter from the secretary of war brought help. Sternberg was assigned the services of a Cpl. Bromfield and six privates, a six-mule team, a wagon with a teamster, and rations for fifty days. In high spirits Sternberg and his entourage set out for the return to the Texas base camp. Along Indian and Coffee Creek, near the Wichita River, he found a fossil-bearing stratum. They stopped the caravan and made camp, staying several days while Charles explored and brought in several fine fossils—a second specimen of *Eryops*, a fin-backed *pelycosaur Edaphosaurus*, and a strange-looking *amphibian Diplocaulus*.[12]

A fire in camp destroyed Sternberg's tent and all his personal belongings. Water thrown on burning sacks of fossils prevented damage to them, but Charles knew it was time to leave. His contract period had almost expired, and he still had work to do—and miles to travel. On April 25, he ordered his crew to pack and move toward Decatur, the railway terminus where he could ship his precious fossils. By May 12, the mission was accomplished and the caravan returned to Fort Sill. From there, Charles headed for Kansas and home.

It would be good to see Anna. She was pregnant again. The loss of the first baby still hurt. He wanted a son to initiate to the wonders of the Sternberg Quarry and the badlands of western Kansas. He had been gone a long time. Now he wanted to work closer to home. After all, he was a family man!

CHAPTER SEVEN

Family Man

Life changed for Charles Sternberg when he returned to Kansas in 1883. Marriage seemed somewhat unreal to him: he and Anna had not yet established anything comparable to a "normal home life." He was, in fact, more of a visitor than a resident in her home.

When Anna gave birth to a boy on August 26, 1883, Charles was overjoyed. He named the baby George Fryer Sternberg to honor his beloved brother and B.F. Fryer, the doctor at Fort Harker, who had worked valiantly to restore Charles to health when he was a teenager. Charles had noble ambitions for this baby. He would be a fossil hunter, of course, and he would be introduced to his inevitable career as soon as he could travel to the chalk beds. Meanwhile, Charles intended to stay as close to Lawrence as possible for several months and made no immediate plans to leave the state of Kansas.

He thought of the Sternberg Quarry, that rich fossil field he had found two years before up near the Nebraska border in Phillips County. That was the most productive bed he could hope to find, and it was only about 250 miles (400 km) from home. He knew he had taken out only a small fraction of the available material from the ravines along the streams flowing through the Solomon River valley that he had explored so carefully.

He spent the winter after his son's birth cleaning and preparing fossils stored in his shop in Lawrence, writing letters to many potential buyers in museums in Europe as well as the United States, and enjoying his little family.

When winter relaxed its hold and travel was safer, Sternberg prepared for another trip to the Sternberg Quarry. This time he was in the employ of O.C. Marsh, who also hired John Bell Hatcher to work with him. Hatcher was young; this would be his first experience on location for Marsh. Later, he would become Marsh's top fossil finder and have a long and successful career developing museums at Yale and Princeton and the Carnegie Museum in Pittsburgh.

Charles quickly established a camp in familiar territory, and was again overwhelmed by the the assortment of fossilized bones—from rhinoceroses, mastodons, and other animals—so accessible in the quarry. The skeletons were mingled with no two bones in natural positions. Hatcher and other authorities believed these bones had been deposited in the floodplain of a running stream, but other experts offered different explanations. Regardless of the original conditions, the mass of bones indicated that after the animals had been overcome by a great flood, subsequent floods scattered their bones. In Sternberg's day, the quarry lay three thousand feet (900 m) above sea level; but it probably was much lower during the lifetime of the species represented in the quarry.[1]

In one area, Sternberg found fourteen feet (4.2 m) of sand and a four-foot (1.2-meter) ledge of rock covering the bones. He had to remove the sand and rock by hand and used a pick, scraper, and plow

Rhinoceros skulls were not infrequent discoveries in the Niobrara chalk of Kansas.

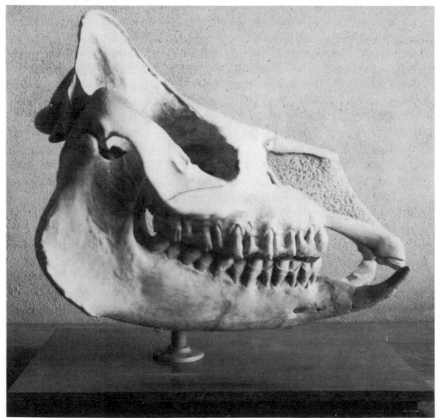

Williston advised Sternberg to build boxes as shown here to ensure safe shipment of fossil bones. Grass or straw was used for packing.

to open an area twenty-five feet (7.5 m) wide and about a hundred feet (30 m) long. Then, resorting to smaller implements, he cleaned the floor and uncovered the bones with oyster knives and other tools he had made to suit his needs. He adapted one such tool from a hoe, rendering it diamond-shaped so it could reach under the high bank to take out specimens that he could not have reached otherwise.

Sternberg's old friend, Dr. Samuel Williston, was also in the area, also working for Marsh. Since their college days and coincidental train trip together in 1877, they had maintained a friendly rivalry as fossil hunters. Sternberg was glad to have him in the area, but perturbed when Williston began to criticize his work methods and packing procedures. Williston felt a responsibility to Marsh and reported that Sternberg and Hatcher were inefficient and wasteful in their work. He told Marsh to urge them to "make strong boxes of moderate size (costing 50 cents each), and in the case of limb bones, to wrap and tie and label each bone separately, then to cover the bottom deeply with hay, put in a layer of bones, filling the interstices carefully and completely with hay, then another good heavy layer of hay, a second layer of bones and remainder with hay."[2]

Sternberg respected Hatcher and "admired his understanding of the work and his thoughtful care."[3] The two men worked harmoniously at first; but soon differences arose, and Williston limited Sternberg to one side of the long ravine and Hatcher to the other. Tempers flared, disputes increased, and correspondence flew to Marsh from both parties. Williston, acting as spokesman for Marsh, tried to mediate and wrote to Marsh of their findings in detail:

> The material is a clean heavy sand and the bones are in remarkable preservation, cleanness and abundance, as an instance it was said not long ago several loads were dug out and sold as old Buffalo bones for $30 per ton [$27 per metric ton]! The

outcrops now, however, have been dug in so far that it is simply folly in removing earth by shovel, that can be done so much more expeditiously by horse-power.

... Sternberg assures me that in travelling this whole region there is no locality that any ways compares with it for abundance, perfection, and cleanness of specimens. The bones have dried, and hardened in the sun can be brushed and cleaned and made to look like the best recent bones. Usually the limb bones are complete and without a fracture, but there appears to be almost no relation between the bones. The animals appear to consist chiefly of Rhinoceros of two kinds, mastodons, large tortoises, and camels and S[ternberg] says also of Carnivores.[4]

Marsh finally made a trip to Kansas himself, leased the quarry land from the owner, and excluded Sternberg from further exploration there, so that he did not return until 1905, long after Marsh's death.[5] Williston resigned from Marsh's employ soon after this episode, expressing his discontent with various controversies and Marsh's methods and attitudes.

Up to this point in his career, Charles had secured work contracts with both Cope and Marsh, depending on the hunting schedule and location. But in 1890, the feud between these two eminent fossil experts would peak. Both men leveled charges against each other for thievery of specimens, espionage, and plagiarism. Both seized every opportunity to ridicule the other's mistakes and character, and both taught their men to be secretive and deceitful about their operations and discoveries.[6] Since this was the heyday of yellow journalism, Editor James Gordon Bennett seized the opportunity to spread the whole scandal across the front page of the *New York Herald* for nearly a week, attracting the interest of masses of people throughout the East who might otherwise never have heard of it. Bennett wrote:

> It is not our business to express an opinion as to the merits of the controversy. Our mission was to gather the news of the impending fray and then to offer our columns as a convenient battle ground on which the gladiators would be insured fair play.
>
> We have only to thank Professor Marsh and Professor Cope for their courtesy in accepting the Herald as the field on which to fight this duel with heavy artillery. We shall leave to the scientists the sad duty of looking after the wounded or burying the dead.[7]

When the feud ended in 1890, Cope's fortune had vanished, partly through some bad investments and partly because of his legal battles with Marsh. Marsh, for his part, had spent much of his inheritance from George Peabody. Both men were now broke, subdued, but unconquered. Cope continued researching at the University of Pennsylvania, but did little more publishing. Marsh continued his paleon-

Charles and family lived on this small farm in Lawrence for thirteen years.

tology work at Yale. The men never made peace with each other, but ceased to feud openly.[8]

Sternberg's sympathies had been with Cope, but he was glad to be half a continent away from the scene of the feud, not directly involved in any court cases with all their bitterness. After being driven out of the Sternberg Quarry by Marsh in 1884, he had returned to Lawrence and dedicated himself to family matters and to marketing fossils stored in his shop.

The next few years proved relatively uneventful for him as a fossil hunter, but rich in the cultivation of family relationships. Anna had given birth to another son, Charles Mortram, in 1885, and to a baby girl, Maud, in 1890. The growing family needed a home, and Charles, with help from his brother George, bought a small farm three and a half miles (5.6 km) southeast of Lawrence. There were fruit trees on the place, and in a short time Charles had a large garden plot established. For the next thirteen years, Anna and the children tended the orchard and garden and packed up the produce, which she drove to Lawrence to sell at the Haskell Indian Institute. It often brought a smile to Charles's lips to think of his wife hauling garden produce to town to feed the Indians. He well remembered his own days of driving horse and wagon to Fort Harker fifteen years before—but not to feed Indians!

Charles took time to visit his parents in Ellsworth. They celebrated their golden wedding anniversary in 1887, with the whole family in attendance—even Dr. George and his wife, who came from Washington, D.C., where he was firmly established as an assistant in the office of the surgeon general. The death of their mother called all the Sternberg children back to Ellsworth again the next year. Charles had reconciled with his father, who, although he never appreciated the profession Charles had chosen, was proud of his achievements and growing reputation and rejoiced that Charles had married and set up a home for his family.

In 1889, a year after his mother's death, Charles decided it was time to take little George with him on an adventure, ostensibly to visit the child's grandfather. It was a good time, he reasoned to himself and Anna, to introduce the six-year-old to the wonderful world of fossil hunting.

Father and son made the trip from Lawrence by buggy with a team of ponies, and the trip provided Charles a means to arouse in his son an interest in the geology of Kansas, just as he himself had seen it some twenty years earlier when he first came to the state. They had a fine time and enjoyed visits with uncles and cousins until one day when, hunting fossil leaves, they left the buggy and horses unattended. Something spooked the horses and they ran—far and fast. Charles and George could only look on helplessly.

Charles took the boy to the home of a nearby relative and set out to catch the runaways, but with no success. After several days, he was forced to return the child by train to his mother in Lawrence. It was a sad ending for little George's first fossil hunt. All he had to show for it was a box full of fossil sassafras leaves and a lively story for his little brother.[9]

Between contracted and sponsored expeditions, Charles continued to collect fossil leaves in and near Ellsworth County. He soon realized, however, that he had supplied most of the museums of the world with examples of the fossil flora of the Dakota formation of western Kansas. He still had about three thousand leaves in storage and no market for them. In his workshop in Lawrence, he cleaned, brushed, and smoothed them with emery stone, and filed them in drawers, hoping to find buyers for them. He used a needle to pry away stone fragments from the leaf petioles, leaving the leaf itself standing up in bold relief.

A Lawrence neighbor, watching Charles at work one day, commented: "You must have taken a long time to carve those things. Why, they look just like leaves!" Charles smiled and offered no explanations, just thanks. He eventually donated most of the specimens to museums across the country. He continued to explore the Kansas chalk beds, but spent much time with his growing family.

By spring of 1891, he welcomed the opportunity to join Williston on a small expedition in familiar territory. Williston, recently appointed associate professor of Geology and Paleontology at the University of Kansas, had arranged a field trip that May to survey the mineral resources of the state. The trip was the first venture for the newly established Geological Survey of Kansas. The party included Sternberg; E. C. Case, Williston's student; and Zoology Professor Vernon L. Kellogg. The group began work at Rush Center, about thirty miles (48 km) south of Hays, where they found Tertiary mammal bones, chiefly rhinoceros. They traveled northwest, following the Smoky Hill River to Elkader, a distance of approximately one hundred miles (160 km). Williston was delighted with what he found and made slides of western Kansas scenery for use in magazine articles he hoped to publish. According to Elizabeth Noble Shor in *Fossils and Flies*, altogether the "collectors found 80 saurians, 35 pterodactyls, 25 turtles, 1 birds, shells, and several rare fishes ... a splendid lot of baculites, barites, and selenite crystals."[11]

Williston realized that one of Sternberg's greatest assets was his uncanny ability to sense where fossils were to be found and to recognize unusual or irregular horizons in geologic formations that suggested presence of fossils. Learning to identify his findings was a slow procedure, but a goal Sternberg strove all his life to attain. He

One of Sternberg's greatest assets was his ability to locate fossil-bearing formations in the chalk beds.

and Williston each recognized the other's greatest assets, and Williston was a patient teacher, appreciative of what Sternberg contributed to an expedition. For Sternberg, this was both an educational and profitable trip.

In the summer of 1892 Sternberg had no contract and no commitment. He decided to head west, back to the familiar and productive territory near Elkader. This time Anna, George, and Charlie accompanied him. Baby Maud stayed in Lawrence with her grandmother. George never forgot that summer, and nearly forty years later, he wrote:

> I took a trip into this region with my mother and brother to join my father ... Little did I realize then that years later I would still be tramping over those old chalk rocks eager to find the slightest trace of a bone or bones which might reward me for my efforts. Perhaps that trip was the one which convinced my father that I should be trained to take up the work he had been following for years. At least I made a discovery which pleased him very much. He has often told me in the years that have passed since, that I was a natural-born fossil hunter.
>
> As I was the oldest son I was allowed to follow Father about over the rough exposures while he searched for some exposed part of a buried skeleton. And one day while he was busily engaged in taking up a specimen he had found, I wandered off on my own account ... Perhaps I was in pursuit of one of the numerous cotton-tail rabbits which would scamper from a shady rock into one of the many holes. Or was I really looking for a fossil? I do not remember. But I have a very vivid impression of coming suddenly upon several large pieces of bones washed down the side of a steep exposure and upon looking up, I saw other parts protruding from a soft ledge above.
>
> I remember running back to my father shouting, "I have found a fossil. I have found a fossil."
>
> ... The thrill of that discovery has never been forgotten. Though I have never since visited the exact spot upon which I made that discovery, I will wager the best hat to be had that I could take you to the very place from which this specimen came.[12]

Charles's mind flashed back to when he was a teenager and hunting fossils brought only disapproval from his father. Charles never swerved from his decision to hunt fossils, and his father remained equally firm in his belief that such activity was only play, not a source of livelihood for a man. Charles resolved to start early in his sons's childhood to show them the excitement found in nature, away from the routine of a "normal home life." As soon as the little boys were old enough to walk, they watched their father clean his fossils and listened to him talk about them.

Charles thought back to 1889 when he had taken George, then age six, on his first trip to hunt fossil leaf impressions. Now he saw his fondest wishes about to come true. George had wanted to find a fossil, and he had found one. Charles envisioned George working with him as a partner in the years ahead. There would be no limit to what he could do with his son as his right-hand man! And there was little Charlie—his time would come, too. Maybe they would be a family of fossil hunters. He liked that idea.

CHAPTER EIGHT

Toil and Triumph in Texas

T he rest of the summer of 1892 was exciting, but somewhat disappointing for young George. He found no more important fossils, though he scanned the rocks daily and lived in expectant assurance that tomorrow he would uncover more great bones to please his father. But it didn't happen. The summer ended, Anna packed up her meager housekeeping supplies, and the family left Elkader to spend the winter in Lawrence. It was time for school for the boys.

Just how long George stayed in school is unknown. He was always a little vague when asked about it in later life. Maybe fifth grade, maybe a little more; it didn't seem really important to anyone except his mother. His father had started fossil hunting while a teenager, although he never ceased to study and learn; and George knew he would be happy and proud to follow in his father's footsteps.

Charles had contracts to go to Texas for Cope in 1895 to further explore the breaks along the Wichita River; and George, not quite twelve years old, was still too young to go along. Anna had given birth to another son, Levi, in 1894, and George stayed home to be "man of the house" and help his mother. Charles hired a farmer, Frank Galyean, as teamster and cook, and they spent much of the next two years working along the Wichita River, returning to Kansas for only short periods, mostly in the winter.

They reached the field area in the summer of 1895. After weeks of disappointing explorations, they found a rich bed of fossil skulls they could not identify and quickly filled a collecting bag with seventy-five pounds (34 kg) of bones, ranging from less than one inch (2.5 cm) to more than eight inches (20 cm) in length. All were species new to

Sternberg, and he was excited, shucking off the despair and discouragement of the previous weeks.[1] Resorting to biblical reference, Charles wrote:

> How can any man who has not had the experience himself, realize the glory of my triumphal march down that rugged trail? Not Nebuchadnezzar, when his chariot headed the army that was carrying away the treasures of the Lord's house from Jerusalem, with the King of Judah, blinded and bound in shackles of brass, in his train, could have known a prouder joy than I did now over this discovery of a new region, in the very heart of the old, which promised so rich a harvest of rare fossil remains. This is an instance of an experience which has been very common in my life—when I have been most completely hopeless and discouraged, I have made my greatest discoveries.[2]

In the three months of the late summer and fall of 1895, Sternberg and Galyean found forty-five complete or nearly complete skulls, many with parts of the skeleton attached; they counted forty-seven fragmentary skulls, ranging in size from less than half an inch (1 cm) to two feet (60 cm) in length. The whole collection contained 183 specimens. Sternberg could not identify them, and the entire collection was sent to the American Museum where it lay undescribed for several months.

Sternberg returned to Kansas late that fall, but came back to the same area January 20, 1896, to continue his work for Cope. Three days later he found a particularly fine specimen of what Sternberg identified as a ladder-spined reptile, *Naosaurus* (also known as *Edaphosaurus*). Charles found the fossil lying in the shale-like flakes of a red and white sandstone. The knobby spines were broken and tilted with the strata and could only be followed with great care. The spines were about three inches (7.5 cm) apart, and Charles numbered them in the order he found them, rather than in the sequence they belonged. He lifted the spine in sections, wrapping about fifty fragments together. These he numbered so that later he could mend and join them. He had found enough of the skeleton for a mounted specimen of the *Naosaurus*, the only such mount in the world at that time.[3]

Sternberg battled not only the elements, but continued discouragement and illness. He suffered recurrent attacks of chills and fever. In desperation he wrote to Cope late in the winter, seeking to be released from his contract and asking permission to go home and rest. Cope's letter of response, dated March 16, 1896, reassured Sternberg of his importance to him and of his "important place in the mechanism of the development of human knowledge." As a result of this encouragement and praise, Sternberg stayed in Texas another month.

In the spring of 1896 Sternberg found a wide expanse of beds that he described as having metallic color, the result, he believed, of a great amount of iron accumulated by an ancient dark, luxurious vegetation. Here he found more fossils in one month than he had found during the rest of the two-year expedition.

One special discovery was a bizarre amphibian that Cope named *Diplocaulus magnicornis*. According to Sternberg's account, it had eyes far down on the face, with two long "horns" that measured fourteen inches (35.5 cm) from tip to tip across. These were prolonged corners of the back of the skull. Cope called these amphibians "mud heads" because they were usually covered with a thin, but tenacious, coating of silicified mud.[4]

Sternberg welcomed the approaching end of his contract with Cope; he eagerly returned to Kansas and his family. His father had died that spring, and he felt a great need to spend time with his wife and children. He made no expedition of consequence the rest of that year, but in the spring of 1897, he returned again to the same area in the Texas Permian for Cope. All specimens were being sent as quickly as possible by express to Cope, and Charles anticipated another good season when, during a violent windstorm in mid-April, while he was wondering how much longer he could last, a courier came to camp with a message from Mrs. Cope, announcing that her husband had died on April 12, 1897.

Even the death of his firstborn son years earlier had not sorrowed Charles more deeply than the death of his friend, whom he described as the greatest naturalist in America. Cope was fifty-seven at the time of his death, and his work was unfinished. Sternberg, always dramatic with words, eulogized: "As long as science lasts and men love to study the animals of the present and of the past, Cope's name and work will be remembered and revered." [5]

The extent of Cope's literary contributions is staggering: a total of 1,395 titles.[6] Many regarded him as the greatest herpetologist America ever produced, and he also attained eminence in geology, ichthyology, mammalogy, paleontology, and evolution. Most of Cope's fossil collection is in the American Museum of Natural History in New York City.

Sternberg had worked for Cope for eight seasons, and he considered the achievements of this association and his work with Cope his most valuable service to science. Sorrowfully Sternberg packed his gear, shipped his fossils, and headed back to Kansas. Young George was now almost fourteen years old. Charlie was nearly twelve, and the baby, Levi, was three.

Charles, believing George was ready to go to work for him, began taking his son to the fossil fields regularly. George had no delusions about the life ahead. He knew it would be rough, hard work, requiring every ounce of strength and energy a man could muster every day. Charles expected George to dig and lift, to climb and walk endless miles across hot barren land. They ate beans, hardtack, and wild berries, and slept wherever they could pitch a tent. The hours were long; the pleasures, few—except those associated with finding a fossil. For a few years they worked in the Niobrara chalk of Kansas, first hunting fossil leaves in Ellsworth County, then moving west to Charles's old base at Buffalo Park and the badlands along the Smoky Hill River in Logan, Gove, and Trego counties.

George soon learned the standard procedure of excavating fossilized bones. When the fossil hunters found a bone protruding from a wash they picked away the surrounding rock until they made a bare rock floor, relatively level, above the bones. Then, stretched out in a prone position and working with an awl and brush, they uncovered the bones sufficiently to determine the position in which they lay. This sometimes took several hours. Then, when the position of the bones was fairly well defined, George helped his father cut trenches around the specimen, chipping down two or three inches (5 or 7.5 cm) on the

A horse-drawn sled brings the huge bundles to the shipping crate and a hoist eases the packing job.

outside of the fossil bones. They built a frame of two-by-four lumber around the specimen, covered the exposed bones with oiled paper, and filled the frame with plaster. Before the plaster could dry, they nailed boards in place on top to ensure a box or panel of even thickness, with all bones in or near their original position.

After the plaster hardened, the strenuous labor of digging around the frame began. The digger had to lie on one side, and chip carefully with a light pick to cut away just enough stone to loosen the frame gently. Using force might have torn the frame and ruined the fossil. With the frame free from its setting, they leveled off the rock and attached the bottom of the frame box. Finally, they packed the boxed chunk into a larger box surrounded by excelsior or other packing material.

When Charles first began hunting fossils more than twenty years earlier, he had tried to remove individual bones, digging patiently with a butcher knife or pick. Then he had packed the loose bones in sacks cushioned with hand-pulled dry buffalo grass. The methods of packing improved through the years, and by 1900 most collectors removed great slabs of rock, rather than individual bones.[7]

In the summer of 1900, Charles and George spent two full weeks in the western Kansas chalk beds, excavating one plesiosaur specimen. The slab was about four inches (10 cm) thick and weighed at least 500 (225 km) pounds. The fossil measured ten feet (3 m) in length, but the slab was much larger and was handled entirely by father and son. The specimen eventually went on display at the Museum of Natural History of the University of Kansas.

Removing and mounting the plesiosaur's bones required the greater part of a year. Williston identified it as *Dolichorhynchops osborni*. Of the specimen's scientific significance, Williston wrote: "Thirty-two species and fifteen genera [of plesiosaurs] have been described from the United States, and in not a single instance has there [previously] been even a considerable part of the skeleton made known."[8]

Among the most common vertebrate fossils found in western Kansas during this period were fragments of the giant fish *Portheus molossus (Xiphactinus audax)*. Until 1900 no complete and perfect example of this species had been discovered. But in the summer of 1901, Charles and George found a specimen of *Portheus molossus* near Elkader in Logan County, Kansas. They raced against the calendar to get it out before winter brought their work to a halt.

It was November, and the ground already frozen hard. Time was precious. The elements had already uncovered the head and trunk of the great fish, but had destroyed portions of the ribs and spine. Cold weather caused the chalk rock to crumble, forcing Charles and George to take up a great slab of plaster with each section. Fighting a strong,

Close up of skull of a large Portheus.

cold wind, they hauled water from a tank one hundred yards away. George, now a spindly lad of seventeen, raced back and forth carrying a bucket of water, splashing its icy contents against his legs as he ran. All the while Charles worked with the plaster and in desperation admonished George repeatedly: "Hurry up! The plaster is hardening!"[9]

Each box of plaster and fossils weighed almost six hundred pounds (270 kg). By the time they had carried them to camp, father and son were exhausted and nearly frozen. The boxes of wet plaster and bones were also frozen and had to be dried out by a slow-burning fire.

Their problems did not end there. Charles and George went home to winter in Lawrence, welcoming the return to comfort and family. Carefully they lifted the heavy box containing the head of their great fish to the table in their Lawrence workshop. But they dropped it. The head was badly shattered, as was the plaster that secured other bones packed in place below the head. Bad luck continued all winter. While the specimen and plaster continued to dry so the fossil could be cleaned and prepared, rats pulled out the excelsior and other packing around the fragile bones, causing them to drop lower and lower in the box. Not to be outwitted by a few rodents, Charles shoved wooden pegs of various lengths under the broken bone fragments, pushing them back into place to preserve the proper angle or position of each bone. He then removed all the packing material and filled the space with plaster.[10]

At long last Charles finished restoring the *Portheus* and sent it to the American Museum for exhibition. Its total length was fifteen feet, eight inches (4.8 cm); length of skull, two feet, two inches (66 cm); and spread of the tail, three feet, nine inches (114 cm). At the time it was first displayed, museum curators called it the most striking example of a fossil fish in any museum in the world.[11]

In 1901, Charles contracted with Dr. Karl Alfred von Zittel of the Bavarian State Collection of Paleontology and Historical Geology in Munich for an expedition to the red beds of Texas. Remembering his difficulties on previous trips when he did not have his own horses, he this time brought his own, and shipped them along with the equipment by rail, with young George in charge. George, now eighteen years old, was a full-fledged assistant to his father. Charles traveled independently, meeting George at Rush Springs in Indian Territory (now Oklahoma). Here he found his son sitting on top of the freight car, waiting and watching, learning everything he could from the brakemen.[12]

Father and son went to an old camp at Willow Springs where Charles had camped on his previous trips for Cope in 1895-97. They

George, about twenty-six years old, in this earliest extant portrait.

91

set up camp in a pasture twenty-five miles (40 km) wide and fifty miles (80 km) long in the area west of Wichita Falls, not far south of the Oklahoma border. Texans remember that summer for its intense heat and drought. Cattle died of dehydration and starvation after becoming so hungry they ate prickly pear, spines and all, and thereby developed infectious sores in their mouths. The mercury often rose as high as 113 degrees Fahrenheit (45 degrees Centigrade) in the shade, and there was little shade. Water dried up in natural and artificial tanks, and the buffalo grass grew brittle and wafted away like dust.

Charles found greater desolation than when he had first been there for Cope. Homesteads, deserted by settlers, had been bought up by cattlemen; even the schoolhouse, once a site for church services, had been moved. Only a few empty houses remained to give testimony to the short-lived settlement of earlier years. As the drought wore on for months, food in the Sternberg camp spoiled. The water, having been carried in a barrel from Seymour twenty miles (32 km) away, was foul—always stale and warm. It seemed that whenever father and son were settling in to a new fossil location, young George would break the bad news that they were out of water. Although the tortuous trip to the well at Seymour was disruptive to their work schedule, it was also restorative: before returning they always drenched themselves at the well, splashing with ecstasy and abandon until their vigor and sanity returned.

Charles not only fought natural forces that summer, but faced manmade obstacles as well. The wagon they had brought from Kansas was narrow gauge, but all the roads in northern Texas were cut by broad-gauge wagons. As a result the team had to pull with one set of wheels in the rut and the other set out. Charles eventually had new axles made. The repairs took a week, during which he recovered from fatigue and a stomach disorder caused by the bad water and intense heat.

Wind, dust, and heat made the idleness of recuperation more unpleasant, so Sternberg gratefully accepted when his friend, Jesse Williamson, offered him the use of a vacant building he owned, situated only a mile (1.6 km) away from the bone bed near Willow Springs. There was a tank of water nearby for the horses and a well only a mile (1.6 km) away. Although only a few buckets of water trickled to the tank each day, it was enough to supply basic needs for the camp. The wait for repairs to the wagon enabled Charles to rebuild his energy level, and after a week, he and George were again tramping in the field with energy and axles restored. For days they found only fragments of skeletons and skulls, frustrating to identify because none of them seemed to belong together.

Tents offered fossil hunters minimal protection from the elements.

Finally they found a nearly perfect skull of a new species, then another skull, nearby. Sternberg did not recognize the first skull, but identified the second as that of a large amphibian with six pairs of large teeth in the roof of the mouth, and a single row of various sizes in the jaw bone. The skull measured twenty inches (51 cm) in length.

Working farther afield but still in the vast pasture west of Wichita Falls, the Sternbergs were again a considerable distance from water. On July 21, 1901, Charles wrote in his daily notes: "It is fearfully hot today, and I cannot work the beds without great suffering " He had to send George to Seymour for feed for the animals and noted, "It is hard on the team to have to haul a load in this weather through dust knee deep, with no water fit to drink."[13]

George returned with fresh water five days later to learn from his father that Dr. Ferdinand Broili, representing the museum in Munich, was coming to visit. Charles rented a large room over a store building in Seymour and built tables where he could arrange a proper display of his findings for the visiting scientist and his party. While preparations were under way, "a storm of grasshoppers struck the building, beating the walls like hailstones; and the next morning the ground was covered with them."[14]

Probably the most prized discoveries Charles made while waiting for the arrival of Broili were two skulls. He identified one, a perfect skull six inches (15 cm) long, as the "batrachian" *Diplocaulus copei* and the other as the iguana-like *Varanosaurus acutirostric*. Many of the bones of the *Varanosaurus acutirostric* had been washed clean by the elements. The upper and lower jaws were locked together, and the long row of teeth shone in the bright light. It was different from anything Sternberg had seen, and when Broili saw it a few days later, he said it was the most perfect specimen ever found in these beds. The *Diplocaulus copei* skull was unusual in that it was compressed laterally instead of vertically. Charles found it among hundreds of fragments of rock filled with glittering scales, as brilliant as when the fish they once covered lived.

George found a bone bed of tiny amphibians known as microsaurs and brought in one skull—the smallest Charles claimed ever to have collected. It had many broken bones and teeth. Charles was immensely flattered later when Broili named one of these minute specimens *Cardicephalus sternbergi*. The skull of this fossil was only one and a half inches (3.8 cm) long; its greatness lay not in size, but in its rarity.[15]

The arrival of the tall, distinguished German on August 8, 1901, erased the agony of the weeks of heat. His praise of the Sternbergs' discoveries and work relieved Charles's exhaustion, anxiety, and uncertainty. This expedition was one of his better paying contracts, and although Sternberg never placed cash value above the scientific worth of his specimens, he had at last begun to realize that even a fossil hunter has to earn a living. He had written to Dr. von Zittel months earlier, explaining that his life had been a constant struggle to secure sufficient funds to carry on his work and that the purchasers of his specimens felt they were doing a good service for museums by securing fossils at the lowest possible price, without taking into consideration the fossil hunter's need for an income. Dr. von Zittel had replied with a generous financial offer, and Sternberg felt a special delight and achievement in this expedition. He was more than anxious to show Broili that the good faith as shown in the correspondence was not ill-placed.[16]

Dr. Broili later introduced a published work on reptiles and on large Permian amphibians called labyrinthodonts by saying of the Sternbergs:

> The excellent results of the expedition of Mr. Sternberg in the spring of 1901 to Texas, which brought many very valuable specimens of *Eryops, Dimetrodon*, and *Labidosaurus* to the Paleontological Museum's Collection, caused the conservator of the Royal Paleontological Collection, Councillor von Zittel, to send out in the year of 1901 a second expedition to the Permian beds of the same territory; he being again successful in securing Mr. Charles Sternberg, the excellent collector from Lawrence, Kansas. Already in June of the same year he was in the midst of his sphere of activity in the Wichita Permian beds, near the small town of Seymour.... On my arrival in the camp, through the assistance of the Royal Bavarian Academy of Science, it was made possible for me to take part in the collection from the beginning to

94

the end of August.... During my stay in that territory, our work principally consisted in making collections from our camp. We were compelled, on account of scarcity of water and from the great heat, to keep near Seymour.[17]

Broili and the Sternbergs had a good relationship that August of 1901. Although Broili's very broken English and Charles's deaf ear often complicated their conversations, George quickly deciphered the accent in Broili's speech and "translated" for his father. Broili spent two weeks in Seymour and out in the fossil field, but had difficulty coping with the extreme heat and dry air. By the end of August the visitor had satisfied himself and his museum administrator and was ready to return home.

Charles and George began wrapping up their work; and in early September, they finally got relief from the heat and drought. Rain fell in torrents for an hour and a half, but within a few hours, the porous soil appeared as dry as before, and work could be resumed. Chill winds of autumn followed closely on the heels of the drought-breaking rain, and the Sternbergs packed their summer's discoveries, gear, and horses and headed back to Lawrence to begin the long process of sorting, cleaning, and preparing the fossils for shipment to Munich.

Charles boarded the train for home and sank comfortably in his seat. How wonderful to relax! His memory scanned the events of the summer. It had been a good expedition—filled with hard work and frustrations, of course, but counterbalanced by the satisfaction of the achievements. Broili was pleased, and von Zittel's praise strengthened Sternberg's self-image. The honor of having his name known and respected and having his fossils displayed in European museums made Sternberg humble. He had hunted fossils for twenty-five years and had worked for some of the greatest paleontologists of his time, but he still suffered from insecurity about his lack of scientific education and inability to recognize and identify his discoveries with authority.

The train chugged through the Indian Territory of present-day Oklahoma, headed for Kansas and home. Charles's thoughts turned to George, a son to be proud of, a son who would follow in his father's footsteps. Right now George was traveling with the horses, equipment and fossils, granting Charles a rare respite from responsibility. What a blessing! And there were two more sons at home!

Quite possibly Sternberg would not hunt fossils again in Texas. After all, he had explored all of the area his employers thought might be productive. Maybe he would find another direction for his efforts; but first he wanted to get to Lawrence—to Anna and the rest of his children.

Charles Sternberg proudly moved his family from their small farm to this larger home in Lawrence.

Piano in Charles Sternberg home in Lawrence, Kansas, 1906.

CHAPTER NINE

God's Great Cemetery

B ack in Kansas, Charles moved his family off the farm and into a permanent home in Lawrence. This remained his headquarters until 1921 when he moved to California in semi-retirement. Charlie was now a teenager and eager to join George and his father in the field. Consequently Charles contented himself for the next several years with further expeditions to the Kansas chalk, usually taking young Charlie, as well as George, with him.

A giant sea turtle, *Protostega gigas*, was the highlight of their 1903 Kansas expedition in the Niobrara Cretaceous formations near Elkader. The skeleton was practically complete, with all bones in or near their original positions. It lay on its back side with forelimbs stretched out at right angles to the median line of the shell. It eventually went to Sternberg's former colleague, Dr. J. B. Hatcher, at the Carnegie Museum of Natural History in Pittsburgh.

Dr. G. R. Wieland, an authority on extinct turtles, wrote in *Memoirs of the Carnegie Museum* that this discovery made possible a full description of the organization of the limbs of the turtle, providing "the most important of the parts yet undescribed as well as the very least likely to be recovered in complete form." Dr. Henry Fairfield Osborn, director of the American Museum of Natural History, also mentioned this turtle in an article in *Science*, commenting on the completeness and size of the skeleton, saying it measured six feet (1.8 m) between the claws of the back feet.[1]

A second and similar specimen, found a short time later northwest of Monument Rocks east of Elkader, was also nearly complete and was taken up in a single slab. It went to W. J. Holland, director of the Carnegie Museum, much to the delight of Charles, who wrote eloquently, "As long as the Carnegie Museum stands, this splendid

example of the great sea-tortoise will be admired by lovers of nature."
Wieland lauded this fossil for the fact that it was complete and had
been taken up in one great slab, permitting a better study of the parts
and their functions. In shape, the tortoise resembled the Mediterra-
nean turtle of today: it had huge front paddles with a span of ten feet
(3 m) and immense claws. The hind paddles stretched out so they
became great oars or miniature boats for the animal. Sternberg
dubbed it "boatman of the Cretaceous."[2]

Sternberg's quest for fossils included not only the remains of large
animals, but sometimes evidence of smaller-scale life forms. Charles
and George also discovered in several localities in southern Logan
County the Cretaceous crinoid *Uintacrinus socialis*. Crinoids are
related to sand dollars and sea urchins and look somewhat like a sea
lily. These tiny animals apparently lived in swarms. The head, or
calyx, of a crinoid had ten long arms as much as thirty inches (76 cm)
in length. A swarm or colony might have as many as forty heads,
flattened out on the underside with all arms intermingled. Frank

*Charles and George found several slabs of crinoids in southern
Logan County.*

Springer, an authority on crinoids, theorized that when death came to one of these colonies, it fell to the bottom of the sea where the first to die were buried in the soft mud and thereby preserved. Others in the colony, not so well protected, disintegrated. As a result, a slab of crinoid fossils will not necessarily show the entire colony.[3]

Charles said that after fifteen years of tramping over this area, he could remember only three localities where these fossils were found, all in the same general area. It was surprising, therefore, when George discovered two more fine specimens about fifty feet (15 m) apart south of Quinter, nearly forty miles (64 km) from the established beds. Charles sold one slab of crinoid fossils to the Senckenberg Museum in Frankfurt, Germany; and Springer bought one. George kept one, which became one of the first displays he contributed twenty-five years later when he became curator of the Sternberg Museum at Fort Hays State University.

George was interested in these tiny crinoids, but he seemed particularly fascinated by mosasaurs, a family of marine reptiles that became extinct near the close of the Cretaceous period. Although their bones are found in many parts of the world, no fossil fields reveal such well-preserved and complete skeletons as those in western Kansas.[4]

George described the mosasaurs as follows: "They looked more like an alligator than anything you could imagine that lives today. Their limbs were flippers, not arms and legs. Likely they never came out on the shore at all. ..." Carnivores, mosasaurs preyed upon the many animals that inhabited the same waters. George found one specimen thirty feet (9 m) long with a four-foot (1.2-meter) long skull—neither the largest nor the smallest known, he said. It, too, is now displayed at the Fort Hays State museum.[5]

Plesiosaurs differed from mosasaurs in that they usually had long necks, short tails, and short, heavy bodies. Charles and George found many examples of both in the fossil-rich beds of Gove and Logan counties. But these were not the only fossil animals they discovered there. Their prizes included: pterodactyls, flying reptiles often mistaken for birds; *Hesperornis*, a flightless bird; mammals; large and small marine turtles; and many species of fish, including *Portheus molossus*, one of the largest scaled fish ever known to have lived.

Charles reverently referred to the rocks of western Kansas as God's great cemetery. For years he had studied the formations of the entire state, from his first trip to Fort Harker a quarter of a century earlier when he and his twin arrived in Kansas, to the innumerable trips to Lawrence in the intervening years. The ancient glacial deposits so familiar in the eastern portion of the state lay deep beneath the surface in Logan, Gove, and other western counties where the limestone formed surface cover for the ancient mysteries far below.

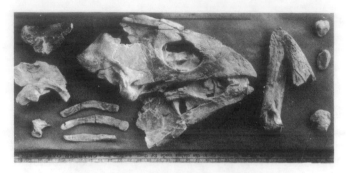

Three sections of bones from a large sea turtle collected by George Sternberg.

Charles, now in his early fifties, realized his lifestyle was wearing down his endurance. He always had to reckon with his stiff leg, deafness in one ear, and susceptibility to ague and other ailments. Nevertheless, he was far from ready to lay down his tools and pass his work on to George and young Charlie. He looked back on great changes and progress and felt no small degree of sadness in the changing tempo of life around him. The Indians were gone; the old scouts were no longer needed. Wagon trains had ceased their slow, rumbling journeys. Instead, mighty steam engines chugged from ocean to ocean, pulling trainloads of humanity, merchandise, and animals. Charles compared the changes in nature and civilization and described the changes thus: "The races of animals, as of men, reach their highest state of development, retrograde, and give place to other races, which, living in the same regions, obey the same laws of progress."[6]

O. C. Marsh had died of pneumonia in 1899. He had spent, according to his own estimate, in excess of $200,000 of his own money on fossils. Having neither wife nor children, he left most of his collection—valued at more than one million dollars—and all personal possessions, including his pretentious home, to Yale College (now University). But at the time of his death his available cash totaled only $186. His fossil collections were so vast that even sixty years after his death, some crates were still unopened. They had accounted for seven freight-car loads for the United States National Museum and twenty-three for Yale. But he had no cash. With Cope and Marsh dead, the old feud of the fossil hunters had become a closed chapter, and interest diminished in fossil hunting in general.

It had been a long time since Marsh had forced Sternberg away from the old quarry near Long Island in Phillips County, Kansas, and now Charles had an irrepressible urge to return to the site—Sternberg Quarry. He knew there were fossils waiting to be found there and figured it would be a fine place to introduce young Charlie to the business of fossil hunting. In the summer of 1905, he, George, and Charlie, now almost twenty years old, made an expedition to the area. The first important discovery was a set of lower jaws of a great mastodon. Although some parts were missing, Charles determined that the jaw measured two feet, six and one-half inches (77 cm) in length, and, at the center of the grinding surface, nine and one-half inches (24 cm) in height. The specimen went to the Royal Museum of Munich.

Nearby, the Sternbergs found numerous chisel-like mastodon tusks scattered among other bones. They also found bones of the three-toed horse, an animal not much larger than a newborn colt of an ordinary horse. Because its toes were spread wide apart, the three-

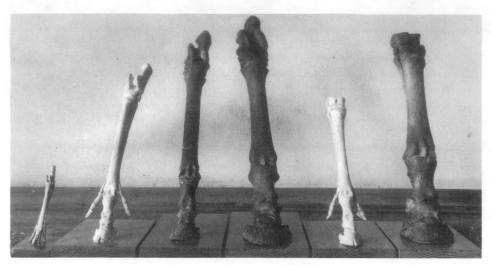

Leg bones of ancient horses.

toed horse could walk over bogs and quagmires, beyond reach of bloodthirsty tigers or other enemies. They found so many loose horse teeth that they deduced that the horses lived in large herds. George developed such a fascination with the remains of the ancient horse that he continued to collect and keep many specimens that showed its development from the tiny *Eohippus* or dawn horse—no bigger than a fox, ten to twelve inches (25 to 30 cm) high—through various stages of development until it evolved into the horse of today.

The Sternbergs also found rhinoceros remains enough for two perfect mounts. One went to the collection in Munich; the other, to Bonn. The summer was a learning experience for Charlie and a productive time for George and his father. They were so busy, in fact, that they spent little time recording the details of their discoveries, contenting themselves with only the bare essentials in their daybooks.

While working in this area with his father, George often went into the little town of Long Island where he met Mabel Smith. Their friendship developed into romance, and they were married December 31, 1907.

George and Mabel established their home on a 160-acre farm southwest of Elkader, not far from the rocks and fossils so dear to George's heart. Here, where George had wandered as a small child and had found his first important fossil, he still loved to tramp among the strange formations of the Monument Rocks and Castle Rock. He knew that each passing year would wash away or blow more topsoil loose and perhaps expose some new and coveted evidence of a great fossil, just waiting for him to find it!

Nevertheless, he and his bride took time for recreation, too. Among his extensive collection of photographs representing about sixty years of fossil hunting are pictures of his adobe or sod house, with a tent pitched nearby, and several wagons and buggies. Several couples are playing horseshoes. Pictures of the "Saturday night crowd" at the Elkader store attest that life was not all hard labor and digging in the rocks. But George's written records described only his business, not his family or personal life and activity.

While George was establishing himself in western Kansas, his father took time out in 1906 to go to New York to attend a gathering of the American Association for the Advancement of Science, meeting in the American Museum of Natural History. Osborn, director of the museum, had become Cope's heir in paleontological achievement, as well as his biographer. The two had been friends since 1877. Snubbed by Marsh in his youth and ordered away from a site because "Professor Marsh" permitted "no amateurs in his digs," Osborn had turned to Cope. The Osborn-Cope relationship in turn opened the door for Sternberg to develop a friendship with Osborn.

Osborn was most gracious to Sternberg, introducing him to a Mr. Hermann, head preparator at the American Museum, who had mounted several Sternberg specimens, and who now escorted the Kansan through the museum's various rooms, including the store-rooms. Emotional and awestruck by the grandeur of this "paradise of ancient animals," Sternberg realized how many specimens were his own and rejoiced at his contribution to the exhibits. "It is a glorious thought to me," he wrote, "that I have lived to see my wildest dreams come true. That I have seen stately halls rise to be graced with many

Horseshoes provided pleasant diversion for the George Sternbergs and friends at their homestead near Elkader.

of the animals of the past that lived in countless thousands, and that I have had the pleasure of securing some of the treasures, in the shape of complete skeletons, which now adorn those halls.[7]

On his return to Kansas he realized his trip East and his expanding activities had attracted the attention of not only the museum at the University of Kansas, but the faculty in the natural sciences as well, and he was occasionally invited to lecture to classes.

Newspapers began carrying stories about Sternberg's activities; and in a long story of Dec. 16, 1906, the *Topeka Daily Capital* attempted to cover his career to date. Charles tried to describe to the reporter the geologic structures of Kansas and at the same time summarize all the ancient life that once populated the prehistoric sea. He loved to refer to earth—especially the rocky formations—as cemeteries of God's dead, with the rich mantle of soil of the short-grass country resting like a blanket over the extinct animals. He referred to crumbling bones of buffalo and flint arrowheads used by the Indians as objects that show how relics of earlier times are preserved.

This feature story and others served to draw attention to Sternberg, and he happily responded to every opportunity to speak or write about his discoveries. He felt smug about his accomplishments and relished every contact he established with men in charge of museums, both stateside and abroad. The period since his most recent expedition to Texas had brought growth to Sternberg's career, not only because of his trip to New York, but also because of his sons' involvement.

Charles wasted no time after returning from New York. In the summer of 1906 he, Charlie, and young Levi, now almost twelve, left Lawrence and headed for western Kansas to meet George. By then Charles had made twenty expeditions to the Niobrara chalk, and he was restless to cover new ground.

That summer they were fairly successful, but found nothing they rated "spectacular." Eventually they found several excellent specimens of Pleistocene age. Quite by chance they discovered a bison with horn cores measuring six feet (1.8 m) from tip to tip and eight feet (2.4 m) along the curve. The Missouri Pacific Railway had cut a new right-of-way across a bend in Sheridan County, near Hoxie, forty miles (64 km) north of the chalk beds. Their excavation came within two feet (60 cm) of the bones buried below, thirty-five (10.5 m) feet from the surface of the earth. The bones were discovered by Frank Lee and Harley Henderson of Hoxie on June 15, 1902, and Sternberg secured them in June, 1908.[8]

Working in unexplored territory in June 1908, the Sternbergs found a great Columbian mammoth, one of the largest of its kind ever discovered, near Ness City, thirty to forty miles (48 to 64 km) southeast of the chalk beds. The greatest circumference of the jaw measured twenty-six and a half inches (67 cm), and it was thirty-two inches (81 cm) long. The last molars pushed out the worn premolars and the other two molars to occupy the entire jaw, with a grinding surface of four by nine inches (10 by 23 cm). The lower parts of the teeth flared out like a fan, measuring twenty inches (51 cm) along the top of the roots.

Moving farther west and north into Logan County, Charles found a nearly complete skeleton of the Cretaceous toothed bird, *Hesperornis regalis*. It lacked a skull, but had a whole framework in normal position with a long neck, a body only about four inches (10 cm) wide, and a backbone like the keel of a boat. Charles named it the "Snake-bird of the 'Niobrara Group.'" It was mounted and sent to the American Museum. George found a skeleton and skull of the flying lizard *Pteranodon* that went to the British Museum.[9]

In evaluating the successes of the previous three summers, Charles felt that the younger sons were now ready to join George and him and to work together on a full-time basis. Levi was 14; Charlie, 23; and George, 25, with no other career in mind than fossil hunting and no intention of striking out independently. They would work as a family of fossil hunters.

George leaves Quinter with a wagon load of supplies.

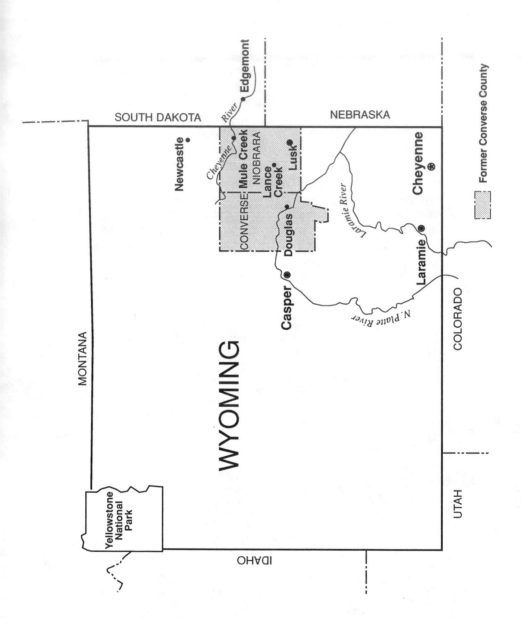

Former Converse County

CHAPTER TEN

Beating the Odds

I n 1908 Charles Sternberg wanted new territory to explore with his sons and a good market for their bounty. He sent out numerous feelers for contracts and sought American and foreign buyers for discoveries he hoped to make. For example, as part of an ongoing sales campaign, he wrote a letter to Dr. Richard Swann Lull listing as available for $300 a slab of crinoids—thirty square feet (2.7 sq m), with 150 calices—discovered by (but not credited to) George the previous year on Beaver Creek, Logan County, Kansas. He also listed two *Portheus molossus* specimens for $350, a 28-inch (71-centimeter) long skull of the mosasaur *Clidastes* for $150, and many other smaller items. In other letters to Lull, he offered such variety as fossil figs from Converse County, Wyoming—thirty for $25—and a seven-foot (2.1 m) *Triceratops* skull from Crooked Creek, Converse County, Wyoming, for $1,000.

He wrote to the British Museum of Natural History, to which he had already sold many fine specimens, and indicated he knew where in Wyoming he could find a good specimen of the ceratopsian three-horned dinosaur *Triceratops*. These dinosaurs had a huge neck frill and weighed about nine tons (8.16 metric tons) with skulls up to seven feet (2.1 m) long.

The British Museum did not offer employment, but responded that it would buy a *Triceratops* specimen if he found one. Sternberg readily agreed.

Sternberg was aware that only thirteen good specimens of *Triceratops* were known to American museums at that time, and half of them were at Yale, collected by J. B. Hatcher, who had worked for Marsh. But many of the fossils Hatcher had found were badly broken and fragmented. Sternberg had seen a map, made by Hatcher, with crosses at sites in Wyoming where noteworthy fossils had been unearthed. Sternberg realized full well that in the past decade the entire area had been scouted inch by inch, and only with good luck would he find something new. Nevertheless, he was determined to launch a family expedition to Converse County, north of the Platte River in east-central Wyoming.[1]

Because dinosaurs have always captivated the imagination of youth, it was only natural that Charles Sternberg's sons—George, Charlie, and Levi—had acquired a deep curiosity about them, hoping one day they and their father could find a "really big one." Levi, at age

Charlie, probably in his early twenties.

fourteen, believed he was ready to follow in the steps of his older brothers and eagerly anticipated the trip to Wyoming. Charlie, now twenty-two, had been working with George and his father since he was fourteen and now was primed to fill an adult role on a major expedition.

Charlie had some specialized duties, as did his brother. He became head cook, responsible for keeping the larder stocked—an important job because the Sternbergs often roamed as much as sixty-five miles (104 km) from their base camp and, intent on the hunt, sometimes failed to notice the depletion of their food supply. Levi started out where Charlie and George had already served their apprenticeship— digging, brushing, carrying water, and helping to lift the heavy frames of bones and plaster. George developed a special aptitude for photography and became adept with his cameras in the field and in the darkroom at home. In the next several years he took hundreds of pictures that proved to be of infinite value in later years.

George had been married less than a year and now moved his pregnant wife, Mabel, to Phillips County, Kansas, to live with her parents in her husband's absence. Like her mother-in-law, Anna

Levi wraps bones with burlap soaked in plaster in Wyoming, 1908.

Sternberg, Mabel learned early in her marriage that Sternberg men lived in a world apart from home. Their wives experienced months of loneliness and separation with slow and irregular communication from their husbands.

Before the Sternberg expedition left Kansas, the family, including numerous relatives from the Ellsworth area, gathered at the family home in Lawrence, fully expecting Charles and his sons to be gone about two years. Actually, they spent three hunting seasons in Wyoming, but broke camp each fall and returned to Kansas for the winter to clean and prepare fossils in the Lawrence workshop.

The Wyoming territory they planned to explore covered approximately a thousand square miles (2590 sq km) in the Cheyenne River country, mostly cattle and sheep range with very little fencing and only an occasional sheepherder for human company.

Having neither firm commitment nor a time schedule, they started early in the spring of 1908 and stopped occasionally during their long trek from Kansas to Wyoming. Shortly before leaving western Kansas, George found a nearly complete skeleton of the great shark *Lamna* with one hundred and fifty long, slender, pointed teeth. More unusual was the presence of almost twenty feet (6 m) of vertebrae. Because sharks have cartilage instead of bones, a skeleton is seldom preserved, although the cartilage sometimes calcifies and, when undisturbed, may resemble a fossilized bone structure.[2] George shipped his fossil to Lawrence, then continued with the plans for the trip northwest.

Florissant, Colorado, high in the Rocky Mountains, was a rich discovery area for gold miners and fossil hunters in its early years.

Many fossils were recovered from this sandstone formation in Converse County, Wyoming.

Charles, young Charlie, and Levi headed directly for Converse (now Niobrara) County, Wyoming. George stopped off in Colorado to explore formations near the South Platte River where he collected fossil leaves, fruit, and flowers from the Tertiary shales. He also found a huge petrified redwood stump and decided to linger a little longer to camp in the mountains with new friends. Harvey Bassler and George Badgley, who had nearby ranch houses, had invited George to visit for a while. However, after a brief stay, George headed north to join the other members of the family near Casper, Wyoming, in an area Charles described as "the border land between the Age of Reptiles and of Mammals, where mammals first appear as small marsupials."[3]

In Wyoming the Sternbergs, hunting day after day with no success, became fearful that they would not be able to honor their commitment to the British Museum. Then, one day in August 1908, Charles and Levi were driving a one-horse buggy to a camp they had made on Schneider Creek. Passing over a small area they had not previously covered, Charles handed the reins to his son and started examining the beds of reddish shale. Here he found the end of a horn core of *Triceratops*, then more of the skull. It was somewhat broken up, and one horn core was missing, but enough of one side and of the back of the head, the crest, and the face was preserved that a restoration was possible. Charles was almost delirious with excitement. After weeks of disappointing search, here at last was the prize he had promised to deliver.

The total length of the skull was six feet six inches (198 cm). The horn core over the eye measured two feet four inches (71 cm) high, and the circumference in the middle was two feet eight inches (81 cm). After excavating it, Charles wasted no time getting his discovery packed and ready for shipment to London. Later he learned to his dismay that Henry F. Osborn at the American Museum of Natural History would have paid him far more for a *Triceratops* than he was to receive from the British Museum, but Sternberg was a man of his word.[4]

With the fossil ready to go, the Sternbergs, now many miles from base camp, were high on excitement, but low on food. Probably they had not reckoned on the power of appetites, whetted by exertion and vast amount of energy expended daily by three young men. Pressured by the family's need for food and the desire to get the *Triceratops* quickly on its way, Charles prepared to take young Charlie and the wagon filled with fossil boxes to the town of Lusk, sixty-five miles (104 km) away, leaving Levi and George to continue work and shift for themselves.

But shortly before Charles and Charlie left the work area, George pointed out a nearby location where he had found a huge bone bed. They explored the area right away, the sons moving in one direction and their father in another. Soon George found bones protruding from a high embankment of sandstone, and Levi found other bones, apparently from the same skeleton, not far away. The prospects were exciting, and Charles was torn between his desire to explore immediately and his need to go to Lusk to get supplies and ship the *Triceratops* bones to the British Museum.[5]

The urgent need for food persuaded Charles to make the trip to Lusk without delay, leaving the explorations to George and Levi. He instructed them to uncover a floor around the exposed bones and see what might be waiting to be discovered. Charlie, worrying about the meager food supply for his brothers, suggested that George ride to a sheep wagon some seven miles (11 km) away and see if he could get some baking powder to make biscuits. After George and Levi watched their father and Charlie leave for Lusk, George mounted his horse and headed for the sheep wagon. He arrived just as the shepherd was preparing to move farther away. Dry weather and lack of water was forcing him to leave—and he, too, was almost out of food. Nevertheless, the herder gave up a bit of baking powder, and George returned to camp that day, only to find they were too short of flour for him to bake anything. Besides they had no salt. George never baked without salt.

George and Levi tried to make their remaining food last; but they soon realized it wouldn't: all that was left was a handful of dried-out potatoes. Hoping for a miracle and the speedy return of their father and brother, George and Levi kept busy digging around the bones. After the first two days, the floor revealed evidence of an entire skeleton, an *Edmontosaurus* (*Anatosaurus*). The boys used tools to follow the line of the bones into the rock bank, starting in the general region of the fossil's hips.

Each bit of dirt they removed revealed more of the skeleton. Altogether, the exposed area they dug measured twelve feet (3.6 m) wide, fifteen feet (4.5 m) long, and ten feet (3 m) deep. By the end of the third day, George could trace the skeleton to the breast bone, having discovered that it lay on its back with the end of the ribs sticking up. Carefully he lifted a large piece of sandstone from atop the breast bone and stared in amazement.

Here was, George realized, "a perfect cast of the skin impression beautifully preserved." Some twenty years later, he wrote of his find:

> Imagine the feeling that crept over me when I realized that here for the first time, a skeleton of a dinosaur had been discovered wrapped in its own skin. That was a sleepless night for me. We were loath to leave our treasure and go for food though we knew there was not a human being for miles to disturb it. Few cattlemen ever rode this way as this was a high and dry region where cattle were seldom found ranging. The sheep men came here with their flocks only when there was melting snow to furnish water.[6]

It was almost dark on the fifth day when the two Sternberg brothers, feeling gnawing pains of hunger after eating only unsalted potatoes for two days, finally saw the horse and buggy approaching. Their first thought was of food, but their father's greeting was "What luck?"[7]

A flash recollection of another fossil find, one that had brought infinite pleasure to his father many years before, raced through George's mind as he remembered the plesiosaur he found when he was a child of nine. George was an adult now, about to become a father himself, and finding fossils was his business. But, a discovery like this was still exciting, and he knew his father would again be pleased. "I have found a fossil!" he proudly shouted, and his eyes twinkled.

Not waiting for Charles to climb out of the buggy and not even reaching for food, George began a detailed account of his find. The brothers had found every bone in place, except for the tail and hind feet, which evidently had been exposed to the air and had washed away. The head lay bent back under the body, and there were traces of "mummified" skin over most of the skeleton. After telling that much, George grabbed a box in the wagon, looking for food. Then he and his father took off for the site. Levi hung back with Charlie, determined to satisfy his hunger as seemed much more important to him at the moment.

It was almost dark by the time father and son reached the floor where the amazing duck-billed dinosaur lay, fully exposed. As long as he lived, George never forgot the great joy he saw in his father's face when he beheld the unique specimen. In his 1930 article, "Thrills of Fossil Hunting," George recorded Charles's reaction: "'George,' Charles exclaimed with a sort of holy wonder in his voice, 'this is a finer fossil than I have ever found. And the only thing which would have given me greater pleasure would have been to have discovered it myself.' "[8]

The dinosaur lay on its back with its front limbs stretching out and up. The hind legs were drawn up against the walls of the abdomen. The Sternbergs theorized that perhaps it had fallen on its back and, because of a broken neck or some other injury, had been unable to draw its head out from under the body. Subsequently it either choked

Ventral view of George and Levi's "mummified" Edmontosaurus.
—Department of Historical Services, American Museum of Natural History, Negative #35607

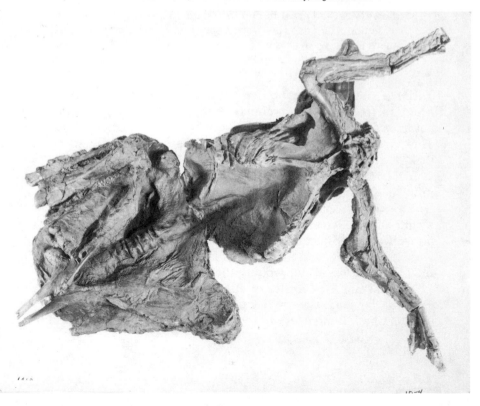

Left side view of the "mummy dinosaur." —Department of Historical Services, American Museum of Natural History, Negative #35606, Anderson

to death or drowned. The fossilized bones were wrapped in the animal's own skin with the polygonal skin pattern plates still showing. The expanded ribs revealed a chest capacity eighteen inches (46 cm) deep, twenty-four inches (61 cm) long, and thirty inches (76 cm) wide. The front limbs showed toes and hoof bones, much smaller than those of the hind legs, and foot bones that had three huge hooves. Alongside the leg bones lay ossified tendons, as thick as a lead pencil in the center and tapered to a small round point.[9]

Charles and his sons slept little that night. Perhaps for George and Levi it was the delicious satisfaction of a full stomach. Maybe it was pure excitement. For Charles, it was sheer ecstasy. He had suffered many disappointments and had come up empty-handed so many times that this discovery swept over him with a wave of thankfulness and new hope. As soon as day broke, he was up figuring out what was needed to get this remarkable specimen ready for shipment.

When Osborn learned of Sternberg's discovery, he immediately dispatched a man to Wyoming to secure the specimen for the American Museum of Natural History, even though he knew that ethically the option for purchase belonged to the British Museum. By appealing to Sternberg's patriotism and promise of a long-running display, he secured the prize. The "mummy dinosaur," unearthed in 1908, is still on display there, lying on its back in a glass case and credited to Charles H. Sternberg.

Of this spectacular discovery, Douglas Preston wrote:

> Lying on its back, with a gaping rib cage and grinning skull ... the specimen looks like a partly decomposed carcass—one can almost smell it—but 65 million years have turned it completely to stone. Technically, therefore, it is not a real mummy, since none of the original body has been preserved, but is actually the fossil of what was once a dinosaur mummy. Not only did the dinosaur mummify under conditions of rapid desiccation and a complete absence of scavengers but once mummified it was quickly covered with sand and fossilized. ...
>
> Skin, tendons, and shreds of flesh—all fossilized—cling to the trachodont's [*Edmontosaurus* = *Anatosaurus*] fossil bones. The animal's head is twisted behind its back in a grotesque arc, which paleontologists believe was caused by the drying and contracting of thick tendons in the neck. The trachodont mummy presents a sharp contrast to the carefully articulated, gleaming black skeletons of other dinosaurs in the [museum] hall, which give an impression of monumentality, stiffness, and formality. Not so much spectacular as it is gruesome, the trachodont reminds us that death was as unpleasant 65 million years ago as it is today.[10]

Closing camp in the fall of 1908 and driving back to Kansas seemed like an endless chore for George Sternberg. He had much to tell Mabel of the summer's achievements, and she had an exciting story for him. Their firstborn arrived October 18 and was named Charles W. Although Charles H. was delighted with his new status as grandfather, family news and activities never really registered with him. His focus was always trained on fossils. He and his three sons were heavily involved with the work at hand. Of course, Levi needed to spend time in school; he had not completed high school. And although George had been able to bypass several years of classroom work, Charles was determined that the two younger boys receive their diplomas.

As he worked in the shop that winter with George and Charlie, Charles mulled over the towering stacks of field notes, day-to-day accounts of his expeditions, and accounts of incidents that would bear telling again and again. After several years of recounting these exploits around the evening campfire in the field, he began to write down the stories and to dream that someday they would reach a wider audience. Gradually that winter he assembled his stories into one manuscript, which he titled *The Life of a Fossil Hunter*. Published in 1909 when its 59-year-old author was the oldest living American fossil hunter, it was the first volume of its kind on the market.

Osborn, who had established the American Museum's Department of Paleontology in 1891 and who had closely followed Charles

Levi provides the scale for a photograph of dinosaur bones found in Converse County, 1908.

Sternberg's career, wrote a gracious introduction, calling fossil hunting a distinctive American profession: "It demands all sorts of hardships of weather, physical conditions, and demands. Hunters must be somewhat of an engineer, have a delicate touch, but strong body to handle the work. Must be content with plain living and find his chief reward in the sense of discovery."

Osborn compared hunters of living game to fossil hunters, saying both are full of excitement, adventure, hope, and anticipation. But the fossil hunter seeks to bring extinct animals back to life instead of bringing live animals close to death. Osborn's closing statement— "His [Sternberg's] is a career full of adventure, of self-sacrifice, worthy of lasting record and recognition by all lovers of nature"— was a glowing tribute to the author.[11]

Published by Henry Holt and Company in a limited edition of five hundred copies, the book was well received. The *Chicago Herald* of March 20, 1909, described it as "a curious union of scientific devotion and religious reverence that is as unusual as it is charming." Boston's *Living Age*, also of March 20, 1909, wrote: "His [Sternberg's] name, as affixed to his specimens, is the only witness to his labors which will remain after him, except the work of three sons whom he has trained to follow in his footsteps; but he has been happy, and his single-hearted story is a book to renew our faith in man's capacity to work for pure delight in work."

The *Topeka Daily Capital* of March 28, 1909, noted: "He [Charles Sternberg] was the man who made fossil hunting a profession. .... *Life of a Fossil Hunter* is the first book ever written dealing almost entirely with fossils, especially fossils of Kansas."[12]

Charles's exploits, of course, did not end with the publishing of his first book. He had worked diligently to help create the Kansas Academy of Science and became a frequent contributor to its published Transactions. His writings appeared in at least a dozen issues, beginning in 1909.

New chapters in the history of fossil hunting were, of course, always waiting to be written. On the agenda for the spring of 1909 was a return to Wyoming to search for more fragments of a major find of the previous summer—another *Triceratops* skull, this one measuring more than five feet (1.5 m) long with horns thirty-three and a half inches (85 cm) in length. The frill had been exposed and was badly shattered. All four Sternbergs had sifted the coarse sand through their fingers, seemingly endlessly and without much success, searching for fragments of the missing frill.[13]

This specimen also had gone to the American Museum in the summer of 1908, and now Dr. Osborn had ordered the Sternbergs back to continue the search for more fragments. After sifting through tons of material, they finally recovered enough pieces to enable the preparators at the museum to make a splendid mount of the skull.[14]

The summer of 1910 was an especially successful one for Charlie in Wyoming. He saw a large part of a dinosaur tail sticking out of a rounded mass of sandstone, and in a ditch below the mass he found another section of the tail. Feeling confident that he had found a worthwhile discovery, he set out in the one-horse buggy to report his find to his father who was camped on the other side of the Cheyenne River. It took most of the day to make the return trip, and by the time Charlie and his father arrived at the find site, it was bitterly cold. They built a big fire of dead cottonwood limbs, and when the last of the fire was gone, they raked the coals away and rolled out their sleeping bags on the warmed earth under a sheltering bank for protection from the biting wind. Levi and George had remained at their father's camp. The next day they broke camp, moving to where Charles and Charlie were working.

The four Sternbergs fought the wind all the next day as they swung their picks for long and hard to uncover a floor on which the skeleton lay stretched out at full length. In looking back over the years of removing large fossils, Charles remembered this as the most difficult of all specimens he ever extracted from the earth. Once, when he

Side view of a Triceratops *skull, found by George Sternberg in 1909 in Converse County.*

Sternberg's Triceratops, *as reconstructed into a standing mount in the American Museum of Natural History in New York.*

flopped carelessly on a ledge of sandstone to rest for a moment, he leaped up quickly, only to suffer a deep cut in his leg and a long tear in his trousers. He had encountered the end bone of the great tail of another duck-billed dinosaur! The huge skeleton lay twisted grotesquely, requiring unusually deep digging to secure all parts of the fossil. The entire body was covered with "skin"—with sand where flesh had been.

The Sternbergs worked two and a half months removing the duck-billed dinosaur, another *Edmontosaurus*, and preparing it for shipment from Edgemont, South Dakota, seventy-five miles (120 km) away from camp. Since it was a most complete and unusual specimen, Charles was determined to save every fragment. The total fossil required three frames—one for each part of the skeleton. One frame weighed thirty-five hundred pounds (1,575 kg), and the Sternbergs had to summon skill and strength to handle it. The skull was more than four feet (1.2 m) long and with the trunk measured more than twelve feet (3.6 m). The tail added five and a half feet (1.65 m). The total weight of the specimen, when removed in plaster frames and boxed for shipment, was approximately five tons (4.5 metric tons).[15]

Charles offered it to the Senckenberg Museum at Frankfurt-am-Main. But a day or two after the museum accepted his offer, the National Museum of Natural Sciences in Ottawa (now the Canadian Museum of Nature) expressed interest in the *Edmontosaurus*, offering almost double the Senckenberg price. Sternberg tried in vain to persuade the administrator of the German museum to give up his claim, or take a substitute, but he was adamant, and the crates were shipped to Frankfurt.[16]

That summer Charles found three more *Triceratops* skulls, and George found an additional one. Two went to the Senckenberg Museum and one to Dr. Marcelin Boule for the Museum National d'Histoire Naturelle in Paris. Charles also discovered a deposit of figs near the *Edmontosaurus* quarry, and five beautiful palmetto palm leaves eighteen inches (46 cm) wide, suggesting a habitat similar to today's Everglades of Florida.[17]

By now George and Mabel had two children, Charles W. and Margery. Mabel brought the children and spent part of the summer at the Wyoming camp on the Cheyenne River. Charles H., generally indifferent to mentioning family life in his writings, took unusual note of this visit: "It was a great comfort and pleasure to have a woman in camp, and we soon noticed a change in the culinary department. It seemed like home to have a daughter and grandchildren in this desert land, and when we came in from a hard day's work in the fossil beds they helped make us forget our labor and our care."[18]

When icy winds began sweeping across the Wyoming plains in late fall 1910, the Sternbergs again broke camp and returned to Kansas, taking with them a load of fossils to be cleaned and prepared. The Wyoming expedition had been exciting. Charles reflected on the warnings he had received when he first decided to go there to look for a *Triceratops* for the British Museum. He knew Hatcher had searched the entire area and found thirteen *Triceratops* skulls and believed he had cleared the area of good fossils. But Sternberg's persistent, untiring efforts yielded six more *Triceratops* skulls and other less significant fossils. Much more important, however, the Sternbergs had found two nearly complete *Edmontosaurus* skeletons, wrapped in their skins—discoveries that would provide paleontologists with an entirely new conception of these dinosaurs.[19]

Charles was tired. He welcomed thoughts of home and Anna, of comfort and relaxation. Pleased with the publication of his *Life of a Fossil Hunter*, he envisioned himself as a public speaker, like his father, who had always been in demand and whose opinions had been highly respected. His brother George was an eloquent speaker with an international reputation as an authority on bacteria and health. A wave of nostalgia swept over Charles when he thought of his doctor brother. He longed to see him again. Maybe that, too, could be arranged, but right now he needed to concentrate on putting his stacks of notes into shape so he could write magazine articles or lecture whenever the occasion arose. He smiled at the prospect.

Charles realized changing family relationships might alter his plans for working together as a family of fossil hunters. George had a family to support; and Charlie, at twenty-five, had found a pretty school teacher who claimed part of his time. Levi, now sixteen, still needed to finish high school. But somehow, Charles knew, it would all work out. Both George and Charlie were firmly entrenched in fossil hunting as a career. Perhaps some arrangements would have to change, but with careful planning surely the four Sternbergs could continue to work together a while longer!

Charles rejoiced at the sight of Lawrence and at the coming of winter.

CHAPTER ELEVEN

New Directions

T he year of 1911 held a strange mixture of tragedy, violence, and happiness for members of the Sternberg family—marked by deaths, natural disaster, birth, marriage, and the initial stages of erosion of the unity father and sons had enjoyed in Wyoming.

George's daughter, Margery, just past her first birthday, died almost immediately after ingesting rat poison she found while exploring the storage cupboards in the family kitchen. His wife, Mabel, was pregnant at the time. With all this, George had his hands full with family problems and responsibilities.

Charles's only daughter, Maud, a young woman of twenty, died of unknown causes, leaving her parents devastated. Charles lectured several times to classes at the University of Kansas, and found solace for the loss of his daughter in writing and producing a small book that he titled *The Story of the Past*. Although she had never gone fossil hunting with her father and brothers, Maud had participated vicariously, listening attentively each winter when the men entertained the family with stories of their expeditions. Charles adored his daughter, and, in his imagination, took her with him on fanciful journeys, escaping reality and floating back through the ages to watch the ancient reptiles, fish and beasts at play. In *The Story of the Past*, he eulogizes Maud by calling her "the joy" of his life, his "comfort and pride" who "came like an angel" in answer to his prayer and to whom he gave his "love" and "care."[1]

Aside from this tribute, *The Story of the Past* has a heavy religious tone and is a telling example of Charles Sternberg's tendency to let his mind wander into a world of fantasy. He recalls all the major incidents of his life from childhood in New York State through forty years of fossil hunting, drops names and places at random, and then abruptly stops. Some have seen a special charm in this little volume and say

Maud, not too long before her death.

Sternberg was ahead of his time, writing delightful fiction with a perception and subject matter not developed by other writers.[2]

Charles spent most of the winter in literary pursuits, but made time to keep tabs on George and Charlie's work in the shop cleaning and preparing of fossils. Levi was busy with school work at Lawrence High School.

One late winter afternoon Charlie looked up from his work on the skull of a *Triceratops* to see heavy clouds amassing. He hurried to put away his tools and get outside the shop to survey the sky, and then ran toward his father's house. Within moments a tornado struck the brick building housing the shop with enough force to topple a wall. The weight of the fallen wall crushed the *Triceratops* skull and collapsed the table on which it had rested, pushing the whole mass through the floor. The skull, intended for the Victoria Memorial Museum Building in Ottawa (site of Canada's National Museum of Natural Sciences), was a severe loss for the Sternbergs, but the fact that no one was injured counterbalanced the loss. Among other fossils in the shop at the time, a nearly complete *Portheus molossus* met with better fate. Rafters from the collapse of the second floor formed a protective shield over the fourteen-foot (4.2-meter) fish, leaving it undamaged although heavily covered with plaster and debris. The fossil eventually went to the Canada's national museum.[3] The workshop, although severely damaged, was eventually restored.

With the approach of spring, the Sternbergs pushed aside their sorrow and misfortune, and planned another summer's work. The repair of the shop could wait until next fall! Charles sent George to

The Sternberg laboratory in Lawrence, Kansas, before the tornado struck, 1911.

After the tornado, some fossils were lost, but others were salvaged, having been protected under falling timbers.

head a small expedition to western Kansas, hoping to find fossils recently exposed by the shifting winds that had swept the plains since Sternberg's last visit there. Such an assignment would enable George to spend more time with his family at a critical period in their lives. In early summer, Mabel would give birth to a healthy little daughter, Ethel.

Charlie looked forward eagerly to returning to Wyoming. He not only had staked his homestead claim on 160 acres (64 ha) of land, but had started construction of a log cabin on the site. He had his own plans for the future. Charles believed that the supply of fossils in Wyoming was inexhaustible and wanted another season of hunting. Levi gladly went along, and Charles picked up a new helper, named Jasperson, in Lawrence before the group headed west.

This time they established camp in the Oligocene formations along Plum Creek, twenty-five miles (40 km) northeast of Lusk. Major discoveries for Charles and his party that summer included a *Triceratops* Jasperson found and a *Titanotherium*, a huge animal somewhat allied to the rhinoceros, that Charles discovered in the Seaman Hills. The hunters searched the floodplain on either side of a dry riverbed south of the Lance Creek beds and reaped a rich harvest of fossil mammals, including the remnants of numerous *Oreodons*, hog-like creatures that once lived in droves. Charles found fifty skulls, and Levi and Jasperson contributed one hundred more to the collection, which eventually went to Ottawa.[4]

Charlie continued to cook for the foursome, but spent much of the early part of the summer completing his house, including one room to become his own workshop. Here he cleaned and prepared Jasperson's *Triceratops*, which was sold to a Paris museum. Here he also worked for several months on his father's *Titanotherium*, a huge specimen that required long and tedious work before it could be shipped. Charlie's summer climaxed on September 18, when he married Myrtle Martin. She was the daughter of a family who lived near Monument Rocks in Gove County, Kansas—"acquaintances" of the Sternbergs. Charlie's father had taken out several fossils from land owned by the Martins, incurring hard feelings over the issue of remuneration. Although it was customary for landowners to welcome the fossil hunters and to give them free rights to whatever fossils they could find, occasionally landowners like the Martins believed they should receive payment just as they would for mineral rights.[5]

Myrtle was a school teacher at the time she fell in love with Charlie Sternberg, and her married life took her far away from her old family home and whatever dispute there might have been. Family photos show Charlie and Myrtle's house as a primitive little cabin in the desolation of eastern Wyoming, but interior views reveal a cozy home,

Charlie, a friend, and Myrtle (right) beside their newly finished log cabin.

Charlie's cabin.

Cabin interior.

a table spread with a white cloth, silver, glass, and china. This was the gathering place for the Sternberg fossil hunters as long as they were in the area, affording them their only touch of domestic comfort.

George had considerable success that summer in the chalk of western Kansas, finding the most complete skeleton of *Portheus molossus* known to science at that time. The fish measured fourteen feet (4.2 m) in length and, when alive, probably weighed somewhere between five hundred and a thousand pounds (225 and 450 kg). It was sent to the British Museum of Natural History. In writing of this specimen in the *London Illustrated News*, a reporter explained:

> It was found at the surface of the ground and was much the worse for the wear and tear of wind and rain and sun. But Professor Sternberg was equal to the occasion. For just as there are two sides to every question, there are two sides to every fossil. The resourceful discoverer determined to get at the other side of this very stale fish; for the exposed side was useless. Accordingly, he covered it with a thick layer of plaster of Paris, and when this was set he proceeded to dig out the fossil from the bed of chalk. This accomplished, he cut away the rock from the specimen, and eventually succeeded in exposing the whole fish, at least the underside.[6]

George's find, like all others up to this date, was credited to Charles H., who was, however, almost apologetic about his thunder stealing in his account of how the specimen was received in England. He realized the time was drawing near when he could no longer lay claim to every specimen located and prepared by his sons.[7]

Charles had promised the *Titanotherium* skeleton to the National Museum of Natural Sciences and agreed that he and George would go to Ottawa and prepare it and a twenty-foot (6-meter) *Platecarpus* as open, or free-standing, mounts. This would be a new experience for both father and son. Heretofore they had done only flat mounts. Now they were to create an open mount, a standing exhibit! They were more than a little apprehensive.[8]

In preparation for this task they decided it was time to make a trip East to visit the museums to which they had sent so many specimens. George had never visited a large museum, and Charles wanted to introduce him to the men whose names were legendary among fossil hunters. They also needed to learn how free-standing mounts were achieved.

They took the train from Lawrence in March 1912 and made their first stop at the museum at the Carnegie Institute and Technical School in Pittsburgh. Of course the paleontological halls were of most interest, and they stood in awe, viewing the great restorations, including J. B. Hatcher's mount of *Diplodocus carnegii*, which had been found in Albany County, Wyoming, and stood twelve and a half feet (3.75 m) high at the hips and seventy-two feet (21.6 m) long.[9]

Charles and George continued on to Washington, D.C., where they visited Dr. George M. Sternberg, the brother Charles had not seen for many years. Charles felt that at last he was in a position to let his brother know he'd made a success of life as a fossil hunter, and attributed no small part of that success to his brother's continued encouragement and assistance.

Dr. George Sternberg now held the rank of brigadier general; he had served as surgeon general of the United States from 1893 to 1902 and as unofficial White House physician for presidents Grover Cleveland and William McKinley before officially retiring.[10] He continued to serve humanity, however, through his creation of the United States Army Medical School, Army Nurse Corps and Army Dental Corps. He worked diligently for several years to establish better living quarters

Dr. George Miller Sternberg. —U.S. Army photo, Kansas State Historical Society

for poorly paid public employees in Washington and continued to study and write about the cause and prevention of infectious diseases.

He won a personal battle with yellow fever, and thereafter dedicated much of his life to the study of bacteriology and sanitation. His reputation as a scientist, researcher, and writer was international, and the library of the Philadelphia College of Physicians cataloged most of his contributions, major and minor—more than eleven hundred titles.[11] He worked closely with Walter Reed on the establishment of the Walter Reed Yellow Fever Commission and gained such esteem that when the Walter Reed Hospital was established in Washington, a major building was named Sternberg Hall.

The doctor had long before given up active interest in fossil hunting, but had kept up with the career of his beloved brother and family. He still cherished the fossils given to him years before in Kansas, as well as the ones he found while he was stationed at Fort Walla Walla, Washington.

Charles gratefully recalled that it was his brother who had first alerted O. C. Marsh, Joseph Leidy, and other paleontologists to the existence of Kansas's vast fossil beds, worthy of exploration.[12] It was his brother who had made possible the first placement of Sternberg fossils in the halls of the Smithsonian. Time and again through the years, Charles had turned to his brother when things were not going well. Dr. George Sternberg had many connections among doctors and scientists, and through them he had been able to help gain recognition for this dear younger brother. Charles's son George F. was his uncle's namesake, and so appreciative was the old doctor of this gesture on the part of his brother, that he remembered young George quite generously in his will.[13]

At the United States National Museum of Natural History in Washington, Charles and George F. visited with C. W. Gilmore, curator of Fossil Reptiles, and J. W. Gidley, curator of Fossil Mammals. They paid particular attention to the preparation of a free-standing mount of the plated dinosaur *Stegosaurus*.[14]

Continuing on to the American Museum of Natural History in New York City, the Sternbergs basked in the glory of seeing their own fine specimens, especially the "mummy" dinosaur fossil. In the museum's Invertebrate Department they viewed the huge *Inoceramus* shell Charles had found 30 years earlier in Kansas. They enjoyed visits with Henry Osborn and with Barnum Brown, his associate paleontologist, who would play a contributing role in the work of the Sternbergs a few years later.

Charles and George went to Yale and marveled at the treasures Marsh had collected. Charles was gratified to find the giant turtle *Archelon ischyros* that he had found years before in Kansas, still considered one of the finest of its kind. Dr. Richard Swann Lull of the Yale museum, a scholar and widely accepted authority on dinosaurs, took the Sternbergs to lunch and then spent some time with them. From this meeting Charles and Lull established a friendship and business relationship that continued for a number of years.[15]

Charles and George took advantage of every opportunity to ask questions, observe and study the technique of preparing a free-standing mount. They found that most fossil specimens are incomplete when the museum receives them, and missing bones may be modeled in clay and cast in plaster. In those days fiberglass and synthetics were unknown. To offset the fossil's brittleness, a framework of iron or steel was needed to support each bone in its proper place. This technique resulted in heavy conspicuous framework, which has long since been refined to make the supports less visible. Vertebrae were strung on a strong iron rod, bent to the position deemed probable. Supports for the leg or limb bones had to be bent to an appropriate position, a matter of guesswork and skill.[16]

Carefully, the Sternbergs studied the process, knowing they would be putting their knowledge to the test as they reached Ottawa in early April. The free-standing mount of the *Titanotherium* was to be accomplished at their own expense, and they had little idea of how great that might be.

In Ottawa Charles and George were somewhat aghast and dismayed when they beheld the room that was to house their exhibit at the national museum building. Boxes and barrels were everywhere, and not a tool was in sight. There was no equipment, nothing to encourage them. Charles took his time, surveying the situation. Then he squared his shoulders, took a deep breath and looked at George. He thought of Charlie in his tiny room of the log cabin in Wyoming with barely enough space to move around the little table that held the skull he was working on. There, with only a few chisels, a knife or two, some brushes and plaster, Charlie had prepared the fine specimen that went to Paris, as well as the *Titanotherium* that Charles and George were now charged to bring to a standing position.

If Charlie could do it, Charles reckoned, so could he and George. Without financial aid and unable to afford a complete workshop, they improvised. Charles wrote the following account:

> For an anvil we secured a disk of solid steel, a strong vise, the necessary half oval, and round steel, and iron tubing for supports, etc. We made a great sandtable first, and laid out on it the skull and [vertebrate] column to get the pose, often getting above it and moving a bone here and there until we were satisfied. We then cut a board so as to fit the contour of the under part of the column, as we had arranged it on the sand table. This board was fastened to

131

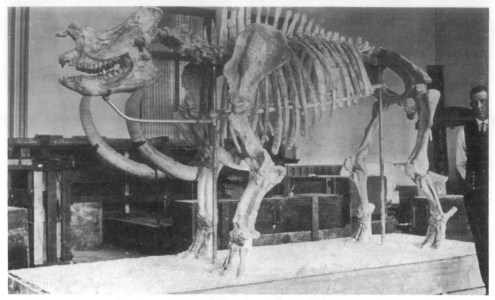

Titanotherium, *the Sternbergs' first attempt at creating an open mount was "made without proper tools," George wrote.*

bases by two halfround pieces of steel that were fastened to either side of the board in pairs, one in front and one behind. These coming together beneath made a round rod of iron that passed into iron tubes a little larger and held them where we wished, with thumb screws. These supports in turn were fastened to broad bases so they would not fall over. We took a cast of the underside of the centre of the vertebrae, and covering the board that served as our model with moulding wax, we stuck the vertebrae in on the central line, giving the exact pose the column had on the sand table. An iron rod was bent so as to pass down the neural canal. The skull, too, was fastened to this iron support, which in turn was fastened to the strong supports that were to secure the skeleton to the base....[17]

By the time the Sternbergs got to the ribs in the reconstruction, they thought the worst was over. It wasn't. The ribs probably gave them the most trouble. They were badly broken; Charles discovered that they would have to drill holes in each end of each fragment, hold them together with iron rods, and secure the whole with some sort of cement. When one kind of cement after another failed, they mixed gum Arabic with dental plaster to create a paste as thick as cream. Then they added LePage's glue and corrosive sublimate to prevent the mixture from spoiling. It was tedious work—and rewarding. They adopted a byword: "We are one rib nearer home."[18]

Several weeks later the task was done. The *Titanotherium* stood—on a cement base. It measured six feet (1.8 m) high at the hips, eleven feet (3.3 m) long to the tail, and four feet (1.2 m) wide at the hips. Although crude by modern standards, the mount, Sternberg judged, "wasn't so bad" for that period, adding that he and George hoped to rectify any "criticism that could be made"—time permitting.[19]

While in Ottawa, Charles finalized a contract of indefinite length with the Geological Survey of Canada, stipulating that he would be its head collector and preparator of vertebrate fossils with the assistance of his sons. They would hunt dinosaurs in the province of Alberta, with the Canadian museums to have first claim on any fossils found. Since work was to begin as soon as possible, Charles and George quickly left Ottawa—Charles heading for Charlie's ranch in Wyoming and George striking out on his own to make arrangements for the next year.

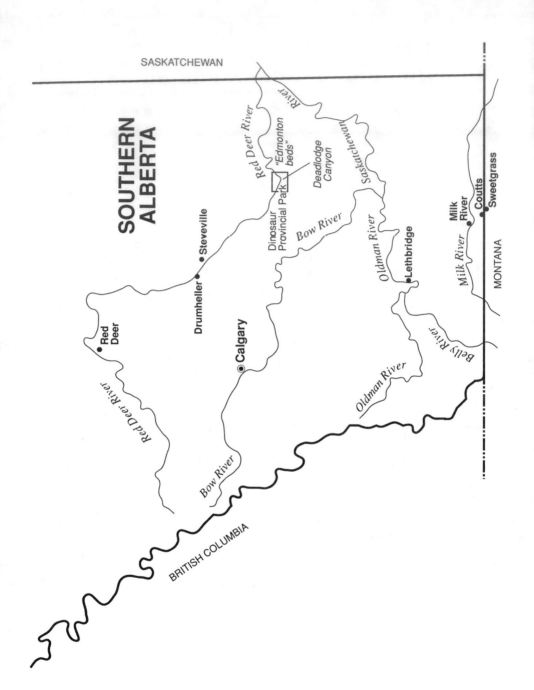

SASKATCHEWAN

SOUTHERN ALBERTA

Red Deer River

"Edmonton beds"

Dinosaur Provincial Park

Deadlodge Canyon

Saskatchewan River

Bow River

Oldman River

Steveville

Drumheller

Red Deer

Red Deer River

Calgary

Bow River

Oldman River

Lethbridge

Belly River

Milk River

Milk River

Coutts

Sweetgrass

MONTANA

BRITISH COLUMBIA

CHAPTER TWELVE

Canada Bound

I t was mid-July when Charles boarded a train in Ottawa, Canada, headed west toward Charlie's ranch in Wyoming where he would meet Charlie and Levi and continue on to the province of Alberta and an exciting, unpredictable future.

Charles leaned back in his seat, watched the landscape, and let his mind wander, to dwell on the events of the previous four months and to anticipate the next several years. He and George had been richly rewarded on their just completed trip. It was quite a thrill to see their own fossils beautifully displayed in the great halls of the museums and quite a privilege to visit with famous scientists and receive assurance that Sternberg fossils were not only acceptable, but desirable. Creating the standing mount of the *Titanotherium* had proved to be more of a challenge than Charles had anticipated, but they had done it! And the Canadians must have been pleased or they would not have offered him the Geological Survey post.

Charles seldom, if ever, rejected a job offer, and usually things turned out well. He hadn't hesitated to sign on with the survey, but now began to second-guess. He had taken a momentous step, and the consequences could be of long duration. In the first place, George was not a part of the contract, at least not this first year. He had made his own deal, obliging him to work with Barnum Brown hunting dinosaurs for the American Museum of Natural History in New York. He would also be in Canada, and, in a sense, competing with his father and brothers. However, George expected to complete his contract with Brown within a year and then to join the rest of the Sternbergs working for the Geological Survey.

Brown, a native Kansan, was some ten years older than George Sternberg. Like George and his father, he had been introduced to fossil hunting early in his youth; but Brown had more formal education than the Sternbergs, having graduated from Lawrence High School and then from the University of Kansas. He spent his summers hunting fossils in Wyoming and while still young, established relations with the American Museum of Natural History in New York, a contact that extended from the day of his graduation from the university until his death.

Brown collected fossils all over the world, but seemed particularly interested in dinosaurs. One of his best known and finest fossil finds is the skeleton of the gigantic carnivore *Tyrannosaurus rex* exhibited at the American Museum. Copies or molds of his specimen are found in major museums all over the country. In Brown's long career of sixty-five years, he is believed to have collected more dinosaurs than any other single individual.[1]

In 1910 Brown made his first expedition to the Red Deer River country for the American Museum to investigate the boastful, but tantalizing claims of an Alberta rancher who had told museum scientists that anyone going to his ranch near Drumheller could find large quantities of bones just like those on exhibit.[2] Brown operated from a large river barge, equipped as a full-fledged base of operations with sleeping tent and galley. This enabled him to float downstream and to carry on board heavy loads of fossils. He returned to the area in 1911 and worked from both the higher "Edmonton beds"—either the Horseshoe Canyon or Scollard formation—and the older and lower Oldman formation. From these two areas, his studies showed evolutionary development among different lines of dinosaurs over several million years of earth history.

Although the Canadian government officially approved of Brown's expeditions, there was some grumbling about what was deemed the ravaging of national paleontological treasures for export. The government had known of the prehistoric remains for more than a quarter of a century, but unfortunately had no one with technical skill and experience to match Brown and his men. Allowed to operate in Canada, Brown planned to move to the lower or Steveville-Deadlodge Canyon area in the summer of 1912, assisted by George Sternberg and others. Brown would continue to work this field until 1915.[3]

The appearance of Charles H. Sternberg was timely for Canadian scientific circles. The scientists figured that this established fossil hunter, with his sons, might be just what the Geological Survey needed to ensure keeping the best dinosaur fossils in Canada and to allow them to compete with Brown. So it was that Dr. R. W. Brock, director of the survey, came to employ Sternberg as collector and preparator.

Two skeletons of dinosaurs were taken from this formation in the Deadlodge Canyon area.

The train puffed and chugged westward and finally, on July 18, pulled into Edgemont, South Dakota, the closest station to Charlie's ranch, not far over the state line in Wyoming. How good it was to set foot on the soil, to stand and stretch and breathe the clean air of the open country! Charles had observed his sixty-second birthday while he was in Ottawa, and sometimes had to admit he felt the changes brought about by age. His stiff leg bothered him more frequently, and his joints hurt in damp weather or when he was cold. Still, he knew he could handle his share of the workload. He wasn't finished yet.

Levi, an uninhibited and bright young man of eighteen, met his father along with a neighbor of Charlie's who had driven a rig from the ranch to Edgemont with big boxes containing the skull of the *Triceratops* Charlie had spent the winter preparing for shipment to the Muséum National d'Histoire Naturelle in Paris. Carefully, the men unloaded the wagon, checked the boxes with the stationmaster, packed Charles's gear and drove back to the ranch. With them was a newcomer, A. E. Easton, a young man from Quinter, Kansas, who wanted to join the Sternbergs as a helper, especially to care for the horses and equipment. Because Quinter was near the Gove County chalk beds, Easton had known of the Sternberg operations there for years. He eagerly anticipated caring for the Sternberg's animals while he learned more about fossil hunting.

Levi about age eighteen.

The next day they drove back to Edgemont and loaded horses and equipment on a freight car with Easton in charge. The Sternbergs boarded the sleeper car and settled down for the long trip to Drumheller, more than eight hundred miles (1,280 km) away. The route was not direct, and the trip would take most of ten days.

Time on the train passed quickly for the first few days. The Sternbergs had never before traveled the route and enjoyed the new scenery—the majestic Rocky Mountains, the picturesque Crow Indian Reservation with clusters of tents. At Great Falls, Montana, they had time to walk across the long bridge and marvel at the Missouri River. Finally, they crossed into Canada at Sweetgrass and Coutts, heading for Calgary.

Levi and Charlie pressed Charles with many questions, and Charles appreciated an audience for his stories about the trip East with George, the wonderful museums, and the difficulties in achieving the standing mount in Ottawa. Neither Charlie nor his father was entirely ignorant about what to expect when they reached the Red Deer River country, having heard about explorations of the area. During the summer of 1874, a group of geologists headed by Dr. George Mercer Dawson, working for Her Majesty's North American Boundary Commission, had found the first dinosaur bones discovered in Canada. They had worked west from the Lake of the Woods in Ontario, north of Minnesota, to western Saskatchewan, north of Montana. Studying the walls of the valleys and the ancient sediments there, they found shattered dinosaur bones weathering out of the canyon walls. They also found shells of extinct marine organisms from the Mesozoic—or Age of the Reptiles.[4]

In 1884 Dawson and Joseph Burr Tyrrell brought out their first notable specimen, the partial skull of the carnivorous dinosaur *Albertosaurus*. It came to the attention of Cope, who studied and described it. Subsequently the discovery attracted enough interest that a party set out for the express purpose of exploring the Red Deer River country for dinosaurs. This party, headed by Thomas Chesmer Weston, found a rich fossil field near Steveville and Deadlodge Canyon. His report suggested that future expeditions use a boat as a base of operations in order to work an area of many miles near the river.[5]

Of course, both Charles and Charlie knew about Barnum Brown's activities and had mixed feelings about George's association with him. They tried to adopt a "wait-and-see" attitude, convinced that George would be able to learn much that would be helpful to their own operation when he returned to the family fold the next year.

Charles knew he was to work under and report regularly to Lawrence Lambe, a vertebrate paleontologist of the Canadian Geological Survey. Lambe, an artist, had considerable experience in studying, describing and interpreting Canadian fossils, but was not particularly adept at collecting specimens. It was Lambe who, through his publications, awakened the scientific world to the significance of the formations along the Red Deer River.[6]

Barnum Brown's camp at the mouth of Sand Creek.

In a neighborly gesture, Barnum Brown helped the Sternbergs "net" themselves as protection from mosquitoes and then took their picture.

The Sternbergs left the train in Calgary to have a rowboat built according to Charles's specifications. From there Charlie and Levi went cross-country to Acme, about thirty-five miles (56 km) north, to meet the train when it arrived with Easton and the equipment. Charles followed as soon as the boat was finished and immediately made the acquaintance of the "pest of the North"—mosquitoes. The Sternbergs soon learned to wear nets when traveling and to keep a smudge going in camp to protect both humans and horses. They established camp three-quarters of a mile (1.2 km) above Drumheller, at that time a small town with only two stores. Within a week they found their first dinosaur bone—from a duck-billed saurian.

The river in the area of the hunt cut a broad, deep valley in places from one to two miles (1.6 to 3.2 km) wide and from four to five hundred feet (120 to 150 m) deep. Tributary creeks had cut narrow trenches farther back into the plain; and long ridges, buttes, pinnacles, and towers transformed the prairie into a veritable picture book of badland scenery. Oyster and clam shells formed the top layer of earth, with bone beds of gray clay, layered with dark shale and seams of coal.

The mixture of clay and sand loosening from the steep cliffs can take on the properties of soft soap when wet and, when mixed with bits of chert, quartz, or other hard substances, can present a genuine hazard to climbers. On more than one wet day Charles slipped and found himself stretched full length on the slope, his face and hands cut by the impact. Wherever the bits of quartz or chert were thick, he could dig in his heels for more secure footing—and invariably find a deposit of scattered bones nearby.[7]

The brightest day of the Sternbergs' first summer in Canada was August 13, 1912, when Charlie discovered a duck-billed dinosaur, thirty-two feet (9.6 m) long and complete, except for the tail. Only the end of the tibia was exposed, but Charlie's sharp eye discerned the irregularity in the formation in which it was embedded. The skeleton was only about one hundred yards (914 m) from the shack of Dan McGee (a local logger and handyman), forty yards (36.6 m) above the forks of the McCheche Creek, and six miles (9.6 km) west of Drumheller. Charlie likened the position of the uncovered specimen to that of a dead dog—pitiful and forlorn.[8]

Because the fossil lay in sandy clay, the bones were cracked into thousands of fragments. To prevent further crumbling and disintegration, the Sternbergs immediately filled all exposed bones with shellac. Carefully they cut away from and under the great mass of bone. They removed the skull only after they had covered it with burlap soaked in plaster. They handled other portions of the specimen the same way—with some bundles weighing approximately three thousand pounds (1350 kg). Six weeks of constant effort finally resulted in the successful removal of the entire skeleton; and as the season ended, the Sternbergs headed for Ottawa with a carload of precious fossil material.[9] The Sternberg wives moved to Ottawa in the fall of 1912 and were able to spend several winters there with their husbands while they cleaned and prepared the fossils they had brought from Alberta.

As the Sternbergs anticipated the 1913 season, they realized the field wagon and rowboat did not meet their transportation requirements: they needed power to navigate the river. Therefore, upon arriving in Drumheller in the spring, they bought a five-horsepower motorboat and then built a barge twelve feet by twelve (3.6 by 3.6 m), large enough to provide a moving base for two tents, placed end to end, one for sleeping and one for a kitchen.

From Drumheller, Levi took the team and wagon and headed south toward Steveville to establish a somewhat permanent camp. Charlie drove the motorboat, and his father and assistant Jack McGee[10] followed on the barge, chugging down the river at five miles (8 km) an hour. The river was about six hundred feet (360 m) wide, and of course,

Levi Sternberg (foreground) takes out the skeleton of a duck-billed dinosaur.

Bones must be wrapped immediately after removal in plaster-soaked burlap.

Cartons of wrapped fossils weigh hundreds of pounds. Horses pull this crate out of the discovery area.

such an outfit was bound to create numerous and exciting—and sometimes dangerous—adventures. Charles recounted one such:

> We were early astir, and Charlie hauled us in mid-stream. A strong east wind blew in our faces. It was disagreeable, because we had to lower our tents to the deck, as they acted as sails, and the power of the wind on them, which was stronger than the current and the five-horse-power motor would have driven us upstream. The choppy waves beat constantly against the front and sides of our scow, curling over the deck itself. The wind howled in the few cottonwoods along the shore and on the islands that we passed. ... About nine o'clock we reached the fifth ferry below Drumheller. The ferry man had stretched a barbed wire across the river; Charlie saw it as he drove his motor under it and shouted to us. Jack McGee, an assistant, rushed for the rear guiding oar and I for the front one. They were both stuck several feet up in the air, and if the wire had caught one, it would have

Charlie Sternberg manages the motor boat pulling the 28-foot (8.4-meter) scow with L. M. Lambe, Charles H. and Levi Sternberg on board.

swamped us. Jack had his back to the wire, and when he released the oar and stood up, it caught his hat and threw it in the river. If the wire had been six inches [15 cm] lower, or the river six inches [15 cm] higher, it would have cut his head off as easily, and thrown it into the river.[11]

Tragedy was averted; the convoy proceeded. After sixteen hours, the fossil hunters reached Steveville, eighty miles (128 km) from Drumheller and near present-day Dinosaur Provincial Park. They quickly realized they had chosen their destination well. Fossil leaves, as well as bones, lay nearby. Except for mosquitoes, the camp proved delightful, and life was fairly comfortable as long as they kept the smudge pots burning and wore netting. Butter, eggs, and chickens were available from a neighboring farmer. A post office was only three miles (4.8 km) away, and good water was accessible. The men counted themselves fortunate indeed as they prepared to enjoy the rest of the summer, exploring the cliffs and badlands of the Oldman formation.

George completed his contract with Barnum Brown and joined his father and brothers in the employ of the Geological Survey of Canada about the time they arrived in Steveville early in the summer of 1913. Very shortly thereafter, Charlie made the first important discovery, a huge carnivore, now known as *Albertosaurus libratus*. As discovered, the double row of ventral ribs, the head and hind limbs, and one foot lay in sight on the slope.[12]

Sensing the enormity and possible significance of his discovery, Charlie took command of the situation and, with McGee as his helper, dug out the fossil segments—aided, of course, by the other Sternbergs. Removal of tons of rock revealed a body so long it had to be removed in two huge sections. The skull alone measured three feet (91 cm) long. Ribs, front and hind legs, and fragments of toes all would be measured later. Now they required separate wrapping and packaging. Excavating, wrapping and packaging them took six weeks of hard labor. Then came the hazardous chore of hauling the sections away from the cliff to camp where the Sternbergs would prepare them for shipment to Ottawa.

The men blasted tons of rock and dumped it on the side of the gulch to make a road. Then they hauled the fossil sections out along a ridge so narrow that if the horses had balked or a wheel had slipped, they would have been dashed to pieces in the gorge below.

Because of the probable value of this exceptional discovery, Charles sought advice about how to mount the specimen—whether he should leave it in its original base and make a panel mount or remove it from all surrounding material and construct a free-standing mount. Gaining permission from the survey director, Charles and Charlie went East almost immediately—to Pittsburgh and the Carnegie Museum, then on to Washington, Yale, and New York, visiting with museum curators and paleontologists who reassured them that the carnivore should be mounted as a slab in bold relief so the bones would stand out clearly, but still be in their original position on the slab of rock.

Levi (left) and helper prepare fossils for shipment.

Back to camp at Steveville they went, where Charlie was to spend the rest of the summer and much of the winter preparing the specimen, proceeding slowly and delicately, as he manipulated his findings. The skull, as noted, was about three feet (91 cm) long, with all teeth in place, each from four to five inches (10 to 13 cm) in length, double edged, with serrated margins. The entire body from front of jaws to end of tail, was twenty-nine feet (8.7 m) long with hind limbs measuring ten and a half feet (3.2 m). The front limbs, in contrast, were only twenty-three inches (58 cm) long. The hind limbs had three huge clawed toes, while the front limbs had only two small and weak claw bones. Sixteen pairs of ventral ribs protected the abdominal walls.[13] The Geological Survey named it *Gorgosaurus libratus*, Latin for "the fierce-looking, easily balanced carnivorous dinosaur." Now known as *Albertosaurus*, Charlie's find would prove to be the most nearly perfect one of its kind known to science at that time.[14]

Each successive discovery meant additional days or weeks of painstaking work to extricate and prepare each fragment. After scraping, cutting, and cleaning each piece, Charlie filled it with diluted shellac or a thin solution of a colorless cement and left it to harden and dry thoroughly. Then he used small chisels, scrapers, or tools handmade by George, who seemed particularly adept at creating them. Since a tool that worked well for one fossil probably would not be as effective on the next one, a continually increasing and varying supply of equipment was necessary. Frequently a bone or portion missing from one specimen could be replaced by some other fragment, cleaned, honed to fit, and carefully cemented to form a reasonable facsimile of the original. Unlimited patience, precision, and caution were necessary, and any slight burst of nerves or impatience could damage or ruin a specimen.

Back on the job later in the summer, Levi found a good specimen of a crested duck-bill lying obliquely across a precipice on a trail over the badlands. The tail was partially exposed from a mass of clay about eighteen feet (5.4 m) high. Charlie then found two more specimens in the same quarry. The men had begun the careful uncovering of the skeleton Levi found, when at the farther end of the discovery, Levi found the tail of yet another duck-bill, leading deeper into the already deep excavation. Since the working season was drawing to a close, the Sternbergs decided to leave the discovery alone for the winter and shoveled tons of earth back over the excavation. They marked the location and waited until the next season.

Temptation was often great to work in haste, to hurry and take out every fossil sighted. But often Charles found haste inadvisable and greediness not to be tolerated. "Preparation of these Red Deer River dinosaurs required courage and patience, not only upon my part, but upon the part of the boys," he wrote. "We must have complete control of our nerves, for a moment's impatience might wreck a specimen we have sought for years. It is a great achievement to mount one of the noble relics of God's creative power in the past. Our laboratory is Holy Ground. The earth is a great plant, from which, for countless millions of years, the Creator has been turning out the creatures of His hand, each 'having seed in itself.' … When, after months of anxiety and labor, we get a specimen mounted permanently for study or exhibition, we are relieved of a strain few can comprehend."[15]

In the summer of 1913 Barnum Brown and the Sternbergs established neighboring camps with easy access to "Happy Jack's" Ferry. There, amid one of the world's richest dinosaur beds, they explored and dug, always on friendly terms, but always with sharp rivalry. "Jack" was a Mr. Jackson, an old cowman who lived in a log cabin near the river. He was range boss for the brother of Admiral Beresford of England, who had developed a ranch. Upon the admiral's death, Jackson took possession of the holdings. He became quite widely known, and the ferry, named in his honor, was an established landmark with an access route across the river.[16]

Both groups of fossil hunters found dinosaurs with skin impressions that summer. The Sternbergs had, of course, enjoyed a similar excitement five years earlier in Wyoming, when they found their first "mummy fossil." For Brown, it was a novel achievement to find such an excellent dinosaur specimen with skin impressions over much of the body.

"Happy Jack" and his cabin.

Also in that one season the Sternbergs collected not only two duck-billed dinosaur skeletons, but other hadrosaurs, ankylosaurs, and carnivorous dinosaurs as well as the spiked-skulled ceratopsian called *Styracosaurus* by Lambe.[17]

Not all discoveries had happy endings. One day that summer Levi found parts of a crested dinosaur—the exposed tail vertebrae, pelvis, hind limbs and many ribs. All four of the Sternbergs joined in the search, but they never found the head, neck, or front limbs. On another occasion, after finding the upper part of an *Albertosaurus* skeleton, Charlie removed tons of rock from where he thought the tail lay, only to discover that it lay in another direction. He never did find the rest of the skeleton.

All expeditions have their serious moments and days, but also lighter sides. Knowing they would be working in close proximity for many weeks, the Brown and Sternberg parties enjoyed a certain camaraderie and often spent leisure hours together, notably on Sunday for picnics. Sternberg photos indicate that Sunday was generally observed as a day of rest and worship, with the men dressed in suits and ties. Charles, true to his religious upbringing, conducted worship services in camp.[18]

Once, late in the summer of 1913, when bull berries were ripe, the Sternberg men picked large quantities, beating them out of the bushes where they grew in protected clusters among strong and sharp thorns. They spread the fruit on tarpaulins on the ground, then loaded the tarps on the boat. Filling pails with berries, they lowered each pail into the clear water of the river to wash away leaves, debris, and

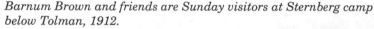
Barnum Brown and friends are Sunday visitors at Sternberg camp below Tolman, 1912.

spoiled fruit, then lifted the red fruit up and out to await cooking on the camp stove. They pressed out the juice, using muslin bags. Then, mixing equal parts of juice and sugar, they made jelly, enough for six gallons for each of the four married men in the party. In camp they used the jelly regularly; it took the place of other fruits and condiments.[19]

George photographed the entire Canadian expedition, taking hundreds of pictures that tell the story not only of the fossils found, but of the struggles of man and beast against the forces of nature—of the flood waters, the cold, winds, storms, the ever-present mosquitoes, and tortuous trips with wagons and sleds pulled by horses up and down the steep canyons and across plains to the shipping point for the heavy boxes of plaster and burlap-encased bones. George also photographed the untamed beauty of the country. All pictures were black and white, and captured the scenic aspects and the lifestyle of the expedition.[20]

A short but alarming incident, captured on film by George, was immortalized in his photo albums with a series of only eight pictures with descriptive captions. The scow had been anchored along shore, and sleeping tents were pitched nearby. The series of photo captions describes the incident:

(1) The river raised very rapidly on June 27th and we were kept very busy clearing away the driftwood from the front of our scow.

(2) As the river raised, large trees and logs were hurled past our scow and we were kept busy pushing them out into the stream away from our boats.

(3) At 6 o'clock June 27th we were compelled to move our sleeping tents aboard the scow as the water began to creep into them.

(4) Thousands of cut logs were hurled past our camp.

(5) On account of high water we were compelled to take our wagon apart and haul it across the river in this way. [Picture shows the disassembled wagon with wheels detached, astride the motor boat.]

(6) Wagon, harness and three men were easily carried across the swollen stream by our motor boat.

(7) At noon on the 28th of June our scow was riding on 6½ feet [1.9 m] of water, directly over where our sleeping tents stood the night before.

(8) The camp here is well out of the current but we were kept on the watch at night for fear the water would fall as fast as it raised and leave our scow high and dry.[21]

Wagon, harness and three men were easily carried across the swollen stream by our motor boat.

At noon on the 28th of June our scow was riding on 6½ feet [1.9 m] of water, directly over where our sleeping tents stood the night before.

Most of the hundreds of pictures that fill two large albums and give a comprehensive story of the years spent along the Belly and Red Deer rivers have no identification other than a film code number. They carry few captions. Apparently, George was so familiar with each picture that no explanation was necessary for him. The albums cover at least three years of the Canadian expedition, with a sprinkling of family pictures taken in Ottawa and Toronto, some laboratory and workshop views of preparatory work on fossils, and dozens of action shots of both the Sternberg and Brown crews at work removing their fossil bundles.

Charles summarized the work of the summer of 1913 in his report to Lambe, dated October 13, 1913, describing their arrival at Steveville on July 1:

> [We] began almost at once to find a rich fauna and collected hundreds of loose bones and teeth in the bone-beds that extended for miles at two horizons through the badlands.... These I did not number, [identifying them] only as 'loose bones and teeth.' They represent the fauna of the Cretaceous during the Belly River [Oldman] time and will be found to contain some valuable material. ... The specimens we secured of greater value, skeletons and parts of skeletons, that we collected and packed in 52 large boxes, are the following...

Sternberg then listed and described thirty items that he considered sufficiently significant to merit individual recognition.[22] He concluded his report with:

> No. 30. — *Large Carnivore*. Parts of maxilla and dentary. Found by Charles H. Sternberg. Loc. Head first lateral ravine to the left that enters the large coulee at "Happy Jack Ferry" in which the road is being built. I was obliged to leave the bones of the skeleton in the rock and brought these parts for identification. We also discovered a number of fine prospects we were unable to take up on account of lateness of season. Charlie has the end of tail of a Trachodon he found in his quarry No. 14 and uncovered several feet of continuous vertebrae under 12 feet [3.6 m] of earth. George found at "Happy Jack's ranch" what promises to be a good skeleton of a Trachodon. I found a good Ornithonimus [sic], etc. but lack of time prevented further work. A storm also was threatened. We reached Brooks, Alta. on the 3rd of October, and by hiring all the men necessary and team we succeeded by working until 9 o'clock P.M. in getting our car loaded, and left for the east at 3:30 A.M. on the 5th, reaching Ottawa on Wednesday, Oct. 8th, when I reported at the Director's Office.
>
> Respectfully submitted.
Victoria Memorial Museum
Ottawa, Oct. 13th, 1913[23]

The Sternbergs seldom left any written record of monetary transactions. However, at the bottom of Charles's report to Lambe he penciled this notation: "35,200 lb. fossils from Brooks, 75 c per hundred, $256.96."

Lambe said in one of his summaries for the survey:

> The principal field work consisted of an expedition to Red Deer River, Alberta, to collect dinosaurian and other vertebrate remains from the Belly River Cretaceous in the neighborhood of, and below Berry Creek [Steveville]. The party was composed of Charles H. Sternberg and three assistants, and its success is to be attributed not only to the skill and experience of those forming the party, but also to the manner in which it was equipped. ... The field collection from these rocks, made by the expedition of 1913, reveals in a striking manner the wonderful variety of the dinosaurian life of the period.[24]

Many years later, Charlie, or Charles M. as he preferred to be called in later life, compiled a survey and map of the area, showing where more than a hundred dinosaur skulls and skeletons had been collected. In 1913 the explorations were still news, and the discoveries exciting. Probably no one at that time realized how rich in fossils the area really was.

Early in the season of 1914 Charles Sternberg and son Charlie accompanied a survey geologist, Dr. D. B. Dowling, to the badlands of the Judith River formation in Montana. Charles, who had been there thirty-eight years earlier with Cope, observed many changes—among them railroads instead of wagon roads and more developed farm lands and irrigation systems. Although he found no vertebrate fossils worth collecting there, he did find bones that proved the presence of hooded duck-billed dinosaurs in the Judith River formation. The three studied the rock structures and compared them to those of the "Edmonton beds" of Alberta and to the beds of Cow Island and the Belly River. Sternberg wrote his summary of the expedition in an article published July 24, 1914, for *Science* magazine, showing no new analyses, but rather reaffirming Cope's report made after his trip to the same area in 1876. Sternberg's talents lay in finding fossils, not in doing research.[25]

Of course the Sternbergs collected invertebrate fossils from all the different strata as they tramped over the large area, but nothing in the two weeks' trip disproved any previous comparisons or changed any interpretations of the similarities or differences.

In July the group returned to the Red Deer River badlands of Alberta. Sternberg recaulked the seams of the scow and moved camp to a new location below the mouth of Sand Creek in the heart of the area known as Deadlodge Canyon. George found another, even finer

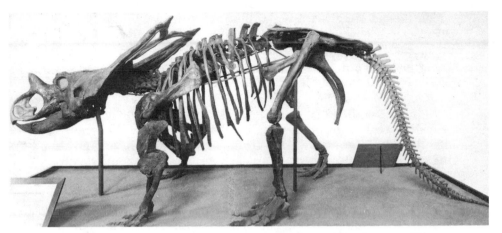

George's Chasmosaurus belli *can be seen in the Canadian Museum of Nature.*

skeleton of the same horned dinosaur, *Chasmosaurus belli*, that he had discovered a year earlier near Steveville. Many years later his two *Chasmosaurus belli* skeletons were mounted side by side in the Canadian Museum of Nature.[26]

Other finds that summer included a well preserved skull of the duck-billed dinosaur *Prosaurolophus maximus*; the skull of another horned dinosaur, *Centrosaurus apertus*; the skeleton of a duck-billed dinosaur lacking the skull but showing excellent skin impressions; the incomplete skeleton of the armored dinosaur *Euoplocephalus tutus*; and the tail club of another armored dinosaur.[27]

At the beginning of the 1915 season George, still working with the Geological Survey of Canada, struck out independently in the Edmonton formation, north of previously explored areas. He went north of Drumheller and west of Rowley and followed the badlands along Battle River, where he worked four months, finding nothing of significance.

Meanwhile, his father and brothers moved their camp south and in June began exploring along the Milk River in southern Alberta. The expedition proved almost fruitless: the best fossiliferous patch of badlands they found was actually on the Montana side of the boundary and therefore out of reach, as far as the expedition was concerned. However, they all had a good chuckle when Charlie was mistaken for a law officer searching for deserters along the International Boundary. War in Europe had tightened inspection of movements along the borders of Canada.[28]

By the end of June the fossil hunters returned to the Red Deer River. Operations were delayed by high water, but eventually they prospected along most of the length of Deadlodge Canyon from One Tree Creek to Jenner Ferry. The most important discoveries were the skeleton of the duck-billed dinosaur *Corythosaurus* and some good armored dinosaur material.[29]

At the end of the working season of 1915, Brown concluded his six-year stint of collecting in the Red Deer River area; the Sternbergs were also planning to wrap things up—to disband their family organization and go their separate ways. George had gradually broken away from the family operation and now wanted complete independence. Charlie had a wife and little son living in Ottawa and tended to draw ever closer to scientific work for the national museums and Geological Survey of Canada. Levi had married Anne Lindblad, whose brother Gustav began working with the Sternbergs in 1915. Levi had no desire to leave Canada, but was ready to stray from the family fold: he had gained respect as a leader of expeditions and as a discoverer in his own right.

Charles was restless. He had not been under a contract to one employer for any great length of time since working with Cope many years earlier. He preferred to free-lance. He enjoyed the public acceptance of his *Life of a Fossil Hunter* and had bared his soul in his small second book, *The Story of the Past*. Now he had accumulated another stack of notes he wanted to organize and publish. He wanted to get started on the next book of the story of his life and decided to call it *Hunting Dinosaurs in the Badlands of the Red Deer River, Alberta, Canada.*

He also felt compelled to set down in print a strange dream that continued to haunt him. One day in the summer of 1915 he had wandered into a coal miner's tunnel in an attempt to escape the midday heat. He found the tunnel floor covered with fine dust that, he found, made a soft bed. He never knew how long he slept, but he dreamed time suddenly flashed back to the era of the Cretaceous ocean. Here—with his wife, sons, grandchildren, and especially his beloved daughter Maud—he watched the activities of an imaginary prehistoric era unfold. So vivid was his fantasy that he made sufficient notes to preserve the tempo, scene, action, and mystery until he could write it up for publication.[30]

Personal desires and circumstances combined to make a decision easy for the Sternbergs. The family of fossil hunters would dissolve its professional ties, but that would not mean an end of fossil hunting for any one of the four.

CHAPTER THIRTEEN

Separate Paths in Canada

O stensibly triggered by a disagreement with Lawrence Lambe over plans for summer field work, Charles's resignation from the Geological Survey in May 1916 was really the culmination of several factors. All the Sternbergs agreed they needed a change from their murky business relationship.

They had never incorporated as a family of fossil hunters, but to paleontologists and museum directors everywhere, theirs had been recognized as a family business, under control and leadership of the father, Charles II. Not only impressed by the Sternbergs' work as fossil hunters, museum preparators found them to be a veritable "spare parts shop" where a museum could order bones of all descriptions to use in preparing restorations of fossils for display.

Although their discoveries were always sold as Charles H. Sternberg fossils, Charles privately gave credit to his sons whenever they made a find. What the sons earned for their efforts is unclear, since the family kept its financial arrangements confidential.

Since coming to the Red Deer River country in 1912, the Sternbergs had found, excavated, and shipped to Ottawa sixteen skeletons or parts of skeletons. And all of these were the subjects of papers and monographs by Lambe, the survey's vertebrate paleontologist.[1]

After his father's resignation, Levi chose to remain with him, free-lancing in the Red Deer River fields southeast of Steveville, working primarily for the British Museum of Natural History. George and Charlie remained, at least for the time, with the Geological Survey.

Charles and Levi secured forty-five large boxes of material that summer, mostly three partial skeletons of *Corythosaurus*, which they hoped to mount "in the rough" as open mounts. In addition, they found a previously unknown turtle and a particularly fine shell of one form already known.

They found a duck-billed dinosaur, nearly thirty feet (9 m) long, a practically complete structure, with some areas of skin impressions. The work of excavation and removal was particularly difficult since the surrounding terrain was inaccessible for horse and wagon. Levi, therefore, walked and carried all the water, plaster, and supplies nearly an eighth of a mile (198 m). Wrapping the specimen was even harder: Levi had to lie on his back, working plaster-soaked strips of burlap under the heavy bones and holding each strip in place until it hardened and stuck to the bone. Then came the task of building a sled road and pulling the heavy load about six miles (9.6 km) in a roundabout direction to the prairie where it could be loaded.[2]

Such labor was not unusual, but in this instance, was all the more frustrating because it was all for naught. The specimen and another duck-bill were loaded on the steamer Mount Temple, headed for Great Britain, but the ship was torpedoed by a German raider, and in a matter of minutes, sank to the floor of the Atlantic. This was one of very few known instances where fossils were victims of war.[3]

For Charles, the duck-bill was truly the "big fish that got away," and with each retelling it became even larger, finer, more valuable and more dearly won.

In 1916, Charles and Levi worked independently in the Deadlodge Canyon area and again had a successful season. They found a medium-sized flesh-eating dinosaur, Albertosaurus, as well as the composite skeleton of the hooded, duck-billed dinosaur Corythosaurus and a skull and part of a skeleton of a small armored Panoplosaurus. Although they offered their finds to the British Museum, the offer was declined, probably because the world was at war and ocean travel was unsafe. The Albertosaurus and Panoplosaurus specimens went instead to the American Museum of Natural History; the Corythosaurus, to the San Diego Museum.

Later in life, when he looked back over his more than fifty years of fossil hunting, Charles always fondly remembered Canada: he reckoned that he had made the most important discovery of his entire career there in 1913 when he found the complete skin impressions of the crested duck-billed dinosaur Lambe called Stephanosaurus marginatus (now known as Lambeosaurus lambe).

Writing for Popular Science Monthly, Sternberg said:

> Prior to this finding there had been much discussion as to the character of the skin which covered the giant frames of these great lizards. In a rock-walled gorge, a mile [1.6 km] wide and five hundred feet [150 m] deep, in Alberta, Canada, I discovered the virtually complete skeleton of this duck-bill, Stephanosaurus marginatus. The small scales of the skin, often mere tubercles, polygonal in shape, were arranged like mosaic in a pavement in

Horses strain and pull as they drag a plaster-wrapped fossil from deep in the rocks of the Red Deer River valley.

parallel rows, along the abdominal walls. Later I found other impressions of the skins of other dinosaurs, and some fragments which are believed to be the actual skin itself, so thoroughly fossilized as to be almost indistinguishable from the impressions of 'molds' in the stone.

This discovery enables us at last to picture with fair accuracy the external appearance of the dinosaurs of these varieties, and, as we know from the skeletal remains their interior arrangements, we now can present the complete duck-billed dinosaur of the crested or crowned species just as it appeared when it strode along the shores of the Cretaceous Sea, or swam in its water.[4]

By the end of the field season of 1917, Charles was ready to leave Canada: the "Golden Era" of dinosaur hunting there was over. The departure of Barnum Brown and the dissolution of the Sternbergs as a fossil-hunting team had simply brought down the curtain on the drama. The valley of the Red Deer River had been a bonanza for dinosaur hunters. Nowhere else in the country were the discoveries so great.[5]

Charles returned to his home base in Lawrence and continued to work independently in Kansas beds for almost four years during which he finished his book, *Hunting Dinosaurs in the Badlands of the Red Deer River, Alberta, Canada*, publishing it himself.

George had begun working as collector and preparator for the Geological Survey of Canada in 1915, concentrating on the "Edmonton beds," north of the area the family had covered in 1912. At the end of

Duck-billed dinosaur material as found in Belly River country by George.

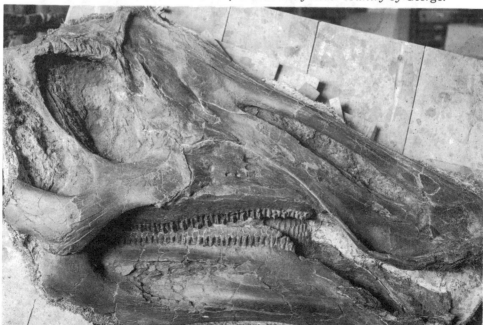

a four-month season he had found nothing spectacular, but several skulls and other materials of some value. The next summer he returned to the "Edmonton beds," working farther south. This time he was more successful, finding the skeleton of the large, flat-headed, duck-billed dinosaur *Edmontosaurus regalis*; the skeleton of a bird-mimic dinosaur, a single specimen or holotype of *Ornithomimus edmonticus*; and also some armored dinosaur material.[6]

In 1918 George resigned from the survey because of lack of work. He left Canada and returned to Kansas to collect in the chalk beds. In 1920 he returned to Canada and the Red Deer River, this time associated with the University of Alberta. He was reasonably success-ful, finding the skull of the hooded duck-billed dinosaur *Corythosaurus*

George searches for fossils in the Mushroom Rocks along the Red Deer River.

excavatus, the skull of the horned dinosaur *Chasmosaurus kaiseni*, and a beautifully preserved skeleton of the soft-shelled turtle *Aspiderites allani*.

At the close of the season Professor John A. Allan of the University of Alberta in Edmonton arranged to purchase the collection for the school's department of geology. He also engaged George to prepare the specimens for exhibit.

George went back again to the same field in the summer of 1921 and collected a nearly complete skeleton of *Corythosaurus* and the skull and incomplete skeleton of the flesh-eating dinosaur *Gorgosaurus libratus*. But the most important specimen was a well-preserved skull and incomplete skeleton of a small dinosaur with an enormously thickened dome-like skull. Dr. Loris Russell wrote of this "dome-headed" discovery: "This thickened mass had been turning up as an isolated fossil for years and had been described by Lambe (*Stegoceras validus*), but now for the first time the appearance and relationships were revealed of this astonishing little dinosaur."[7]

George later described the discovery:

> The first week in the field found me going over some very promising sandstone exposures when by the merest chance, I saw three small teeth glistening in the sunlight, two pointing one way and the other meeting them. I did not recognize them at all. They were small but perfectly preserved. I soon had a floor laid bare and in less than an hour I knew I had a perfect skull of this little animal as well as most of the skeleton. I had made a very important discovery. I knew that when this specimen which I had just uncovered was described it would set at rest much uneasiness and uncertainty regarding this very odd, though small animal.
>
> I went back to camp taking the skull with me. I dug up every bit of literature dealing with it that I possessed, for there had been much written discussion among vertebrate paleontologists about this queer animal. I wrote a long letter to the University at Edmonton, telling them all about my find. I was thrilled to the very marrow to think that at last I had been the fortunate discoverer of this little known animal. ... The greatest thrill of all came when in 1924 I received a bulletin from the University of Alberta with a complete description of this fine specimen which is housed in the museum of that great University. Mr. C. W. Gilmore of the United States National Museum at Washington, D. C. had been the man chosen to study the specimen. He recognized it as belonging to an animal named *Troodon*, described from a single tooth by Leidy back in 1856. He also recognized that it introduced to science a new family of dinosaurs entirely different from anything with which it had previously been associated.[8]

The discovery of this fossil had been photographed by a motion picture crew—much to the surprise of George at the time. The Dominion Motion Picture Bureau, intending to capture the work of the dinosaur-related efforts of the Geological Survey of Canada, had gotten its directions confused; and its crew unintentionally visited the George Sternberg camp instead of the Geological Survey's camp, then under the direction of brother Charles M. As a result, the film shows the work of the University of Alberta, both in the field and at the national museum in Ottawa. It traces the operation through the exploration, discovery, removal, preparation and exhibition of fossils.[9] Some of the early footage has been incorporated into a more recent film for the National Film Board about the life and work of Charles M. Sternberg, called simply, "Charlie."[10]

George spent the winter of 1921-22 working on the preparation of his specimens for the University of Alberta, then in the spring he accepted a position with the Field Museum of Natural History in Chicago. This assignment began with work in Canada, but soon took him thousands of miles away to South America.

George finds a dinosaur bone protruding from a rocky exposure, 1922, shortly before leaving for South America.

Thirteen years later he returned to Canada temporarily to complete some unfinished preparations he had abandoned earlier.

Levi, the youngest son, was the most spirited of the three brothers and generally conceded to be the prankster. His sense of humor and ability to shrug off difficulties lightened family tensions and endeared him to his brothers. He made some important discoveries while working with his father, but he attained his greatest stature after his father left Canada. Like his older brothers, he was introduced to the world of fossil hunting at a tender age. He was only fourteen when he went with his father and brothers to Wyoming in 1908 and was still a teenager when the decision was made to move to Canada. He usually shouldered his share of the menial tasks—shoveling rocks, digging, scraping, lifting, packing, and hauling, but he seldom took part in the preparation of the specimens, once they were in the laboratories. He loved outdoor work, was indifferent to study and scientific investigation, and happy with his lot in life.

Levi and his wife Anne established their home in Toronto and enjoyed a long life together, despite his frequent and prolonged expeditions to the fossil beds a thousand or more miles away.

Levi continued to work with his father after the older brothers pulled away from the family operations in Canada. But when Charles H. decided to return to the United States, Levi looked elsewhere for employment and was fortunate in being available when and where his expertise was needed.

The University of Toronto had begun assembling a paleontological collection soon after the university was established, and by 1908 the distinguished zoologist Professor B. A. Bensley had visited the Red Deer River badlands and been able to acquire what was deemed a "fairly good collection." In 1910 the university had purchased the skeleton of the mosasaur *Platycarpus coryphaeus* from Charles H. Sternberg. The university struggled for several years to discover and prepare specimens from the Red Deer River valley, but it lacked anyone with knowledge of the area and experience in taking out and preparing the specimens.

Finding Levi Sternberg was lucky for the university. In late 1919 he was engaged as head collector and preparator for the Royal Ontario Museum, working under Dr. William Arthur Parks. It was too late to do much field work, but he and Parks made ready for the new season by setting up separate camps five miles (8 km) below Little Sandhill Creek near Deadlodge Canyon. They each had an assistant and worked independently. Both teams were fairly successful; and that same winter Levi, assisted by Robert Wilson, who was Parks's right hand man, completed the preparation and mounting of a skeleton Parks had obtained in 1918. Good publicity and placement of this choice addition to the museum made it easier for Sternberg and Parks to gain support for further acquisitions.[11]

Wilson left the Royal Ontario Museum staff in 1920, and Levi brought in his brother-in-law, Gustav Lindblad, who had worked with the family group for the previous four years. Sternberg was now in charge of the expedition. They went back to the Little Sandhill Creek area and had a successful summer, finding "a nearly complete skeleton of a duck-billed dinosaur in which the hood was drawn out as a great scimitar-like crest extending back over the neck and shoulders, a type of *Parasaurophus walkeri*."[12]

Levi and his men enjoyed increasing success in discovery and interest in the growing museum and seemed settled for life. In 1925 he led an expedition to a Pleistocene locality in Saskatchewan, and then returned with his staff to the Red Deer River badlands, near Steveville. Among his discoveries in the vicinity of Steveville was an incomplete bird-mimic dinosaur, identified as a type of *Struthiomimus samueli*. This specimen was important because the skull was nearly complete and revealed for the first time the cranial structure of ornithomimic dinosaurs. The field party also collected the skull of a *Lambeosaurus* and three nearly complete skeletons of cerotopsian dinosaurs. One of these was in especially good condition and was prepared as a free-standing mount. Another was later discovered to be a new species, a holotype of *Chasmosaurus breviorostris*.

By 1927 Levi and his museum staff began exchanging specimens with distant museums, sending, for example, a duck-billed dinosaur to the Los Angeles County Museum of Natural History. The staff spent two field seasons in the Tertiary rocks of western Nebraska and

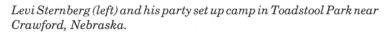

Levi Sternberg (left) and his party set up camp in Toadstool Park near Crawford, Nebraska.

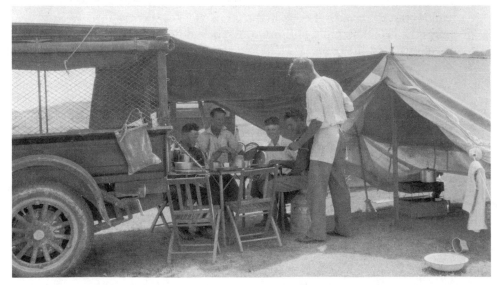

Levi and party atop the strange formations in Toadstool Park, northwest of Crawford, Nebraska.

Wyoming, but of course, the greater part of their finds were in western Canada—the "Edmonton beds," the Steveville area, and Little Sandhill Creek.[13]

In 1935 the Royal Ontario Museum party made an unusual discovery several miles east of Little Sandhill Creek. It was indeed familiar country to Levi Sternberg. He had been here with his father and brothers in 1913 and 1914, when they found the skull of the spike-frilled horned dinosaur *Styracosaurus albertensis*, which went on to win a prominent display in the National Museum of Canada. Now, back in the same area, Levi noted that in the twenty-two intervening years the creek bed had shifted and exposed additional bones.

As Levi and his party worked feverishly, but carefully, to uncover a find in this familiar territory, Levi realized he was uncovering more of the *Styracosaurus albertensis* that George had discovered and that he himself had helped remove long years before. He found the lower jaw, for which the Sternbergs had searched diligently in 1913. Now it lay almost in plain sight where the small stream had shifted its course. He continued searching and recovered most of the rest of the skeleton—in scattered and fragmented pieces. It took some years, however, before this material was finally reunited with the skull in what is now called the Canadian Museum of Nature in Ottawa. Only the skull is currently on display.[14]

In the late 1950s, Levi, with Ralph R. Hornell and Allan Weare, reopened the bone bed in the Upper Milk River Valley; but the discoveries were disappointing, and the group moved on to the Oldman formation near Lost River. Still not finding anything worthy of collecting, they moved again, but again had only minimal success.

Levi continued to work with the Royal Ontario Museum until his retirement in 1962 with the rank of associate curator. He had led twenty-one expeditions to various fossil fields, mostly in western Canada and northern United States. In his later years he perfected several latex casting techniques, both for replication of fossils and for making flexible casts of modern fishes and reptiles for display.[15]

Levi was considered by paleontologists to be a meticulous collector, with an amazing ability to judge the condition and orientation of a specimen, after only a few minutes of work with a pick. He died at the age of eighty-two in Toronto on October 21, 1976, his wife having preceded him in death by a few months. They had no children.

Dr. Loris Russell, curator emeritus of the Royal Ontario Museum, wrote of Levi: "In the field he was a hard worker and a congenial companion, always joking and teasing. He always enjoyed excellent relationships with local people wherever he was working. Levi will be remembered as one of the last of the old-time bone-hunters, as a personal friend and as one who contributed so much to the Vertebrate Paleontology collections at the R.O.M."[16]

Levi Sternberg is in good spirits as he finishes cleaning up a fossil, using a small whisk broom to wipe away loose waste material before lifting his prize and wrapping it in plaster. —Royal Ontario Museum, Toronto, Canada

CHAPTER FOURTEEN

A Namesake Makes a Name for Himself

The day he found his first fossil was not etched in red in the calendar of Charlie Sternberg's memory. Neither could he pinpoint any pivotal personal experience that might have persuaded him to spend his life with fossils. His career decision just happened, sometime in childhood; and he never wavered for the rest of his life—except for one brief period when he toyed with the idea of being a teacher, like his grandfather. Charlie did have dim memories of the summer of 1892, George's momentous summer. Not quite seven years old, Charlie had gone with his mother and brother to western Kansas to spend the summer with his father. They had lived in a tent near the rocks where his father worked and where George found his first fossil. Somehow, George's first fossil find seemed to seal the future for both Sternberg brothers. Their father expected them to be fossil hunters, and good children tried to live up to parental expectations.

Although named for his father, Charlie took more closely after grandfather Levi—the educator, writer, scholar, and minister. Charlie also admired the illustrious career of his father's famous brother, George M. Sternberg, who was a gifted writer, as well as a doctor and researcher. Charlie saw fossil hunting as a way of life, but also as a means to an end—a learning experience that eventually opened broad vistas for him, and opportunities for research and publishing. Fortunately, he was able to interweave several careers during his long life.

The dissolution of the family business in 1916 allowed him to step out from his father's shadow and into his own light. He had reaped assorted benefits from the family business and the contacts it afforded, but he was ready to move toward more scholarly endeavors.

His interest went beyond finding and removing the evidence of prehistoric animal life. He wanted to know more about each fossil and to understand more about the world as it was eons ago.

He cared nothing about writing of his own life or discoveries and adventures. He wanted to study and write about the fossils themselves.

As early as 1911, when Charles H. was basking in the popularity of his first book, *Life of a Fossil Hunter*, Charles M. saw his first article published: "New Model of a Hooded Duckbilled Dinosaur" appeared in *Geological Society of America Bulletin 52*. The younger Charles Sternberg was called "Charlie" by his friends all of his life, but when he first began writing for publication, he chose to use the more dignified Charles and added the "M." to avoid confusion with his father. Ever since he was a child, he had to contend with having two Levis, two Georges, and two Charleses in the family.

He read voraciously. He listened and observed, and whenever he could find time, he put his thoughts on paper. In 1917 his article, "Notes on the Feeding Habits of Two Salamanders in Captivity," appeared in the *Ottawa Naturalist* 30. In 1920 he began contributing somewhat regularly to scientific journals in both Canada and United States. His bibliography, as compiled by Richard Gordon Day, lists sixty-five titles of articles written between 1917 and 1970. The *Canadian Field Naturalist* and the *Geological Survey of Canada* published the bulk of his material with such titles as "A Popular Description of Dinosaurs," "The Bison and Its Relations," and "The Skull of *Leidyosuchus canadensis lambe*." Among his last articles were "White Whale and Other Pleistocene Fossils from the Ottawa Valley," published in 1952 in *National Museum of Canada Bulletin 123*; and "Comments on Dinosaurian Preservation in the Cretaceous of Alberta and Wyoming," published in 1970 by the *National Museum of Canada, Publication in Paleontology*[1]

Charlie not only made remarkable fossil discoveries, but he also recognized the biological and geological significance of them. "He became a competent anatomist, a discerning biostratigrapher, and a creative interpreter of the adaptations and mode of life of the animals whose remains he discovered," wrote Dr. Loris Russell, who worked with all the Sternberg men at various times in the field and later served as curator of the Royal Ontario Museum.[2]

Charlie was known as a finder of spectacular fossils. His father proudly wrote of his discovery in 1909 in Kansas of the skeleton of the toothed bird *Hesperonis regalis*, sometimes called the Royal Bird of the West. The skull was missing, but the rest of the skeleton lay in normal position with a narrow body, only about four inches (10 cm) wide. This fossil went to the American Museum of Natural History.

The next year in Wyoming, in addition to other fossils, Charlie found a remarkable duck-billed dinosaur, unusual because the entire body was covered with skin. This specimen had gone to Germany. Other discoveries followed one after another, and all were recorded by his father in the accounts of the family operations.

Discovery of a fossil was exciting for each hunter, and the hard labor of removal or recovery exacted every ounce of strength and patience he could muster. But acquisition was neither the ultimate achievement nor the total satisfaction for Charlie, whose concern did not stop with the discovery and excavation of fossils. Russell described him as "a fine preparator, meticulous, and competent." Exceptionally patient and skillful, Charlie not only was concerned with how each specimen should be mounted, but also wanted to understand the distribution and life habits of the ancient animals. He became a creative interpreter of the adaptations of the prehistoric life he found.[3] At the same time, he continued his exploratory work, and for twenty-six years he combined two different, but related, careers.

After the departure of the other members of his family from the employ of the Canada Geological Survey, Charlie continued to work in Ottawa collecting and preparing fossils under direction of Lawrence Lambe. When Lambe died in 1919, he left unfinished a paper on a skeleton of the armored dinosaur *Panoplosaurus*. Encouraged by the chief paleontologist, E.M. Kindle, who assisted him, Charlie prepared a supplement to Lambe's unfinished manuscript, and thus began in earnest his long and productive career as a scientific writer.[4]

In 1921 Charlie visited the Morgan Creek badlands southwest of Wood Mountain, Saskatchewan, where George Dawson had found the first Canadian dinosaurs in 1874. After discovering several specimens there, Charlie was able to show that the Saskatchewan dinosaurs dated from the latest stage of the Cretaceous period, from the beds then known as Lower Ravenscrag (later renamed the Frenchman formation).[5]

During that same time, working mostly in the Oldman formation of Alberta, Charlie continued to hunt, find, and describe skeletons of dinosaurs and other vertebrates. He found items in areas that had been searched several times before, but he also sought out new beds. In collaboration with F. H. McLearn, who was mapping a region in southern Saskatchewan in 1928, Sternberg found *Triceratops* and other late Cretaceous dinosaurs in the Frenchman formation in the Eastend area, and found Miocene mammals in the Wood Mountain gravel.[6]

Charlie led thirteen expeditions to the fossil fields of western and eastern Canada, and many of the specimens he added to the collection of the Geological Survey and the national museum in Ottawa were new to science. He made these specimens the subject of anatomical description and systematic analysis in a series of papers in which he defined seventeen new taxa, and proposed several hypotheses to account for the distribution and the life habits of various dinosaurs.[7]

In 1928 Charlie returned to the Red Deer valley, near the Steveville Ferry where he had launched his Canadian exploits years earlier. In the intervening years, other hunters had crossed and recrossed the area, finding an occasional fossil; but, by carefully covering the area on this trip, Sternberg and his party were able to find a number of important specimens, including the skeleton of the hooded duck-billed dinosaur *Lambeosaurus clavinatalis*, the skull of the small hooded duck-billed dinosaur *Tetragonosaurus cranibrevis*, and part of the skeleton of the horned dinosaur *Chasmosaurus russelli*. Also, in a small area south of the Steveville ferry, he obtained the hind feet of two small flesh-eating dinosaurs, *Macrophalangia canadensis* and *Stenonychosaurus inequalis*.[8]

Charlie's field of study ranged from Nova Scotia to British Columbia, with some of his most important discoveries made in the western provinces. A trip to Peace River Canyon in British Columbia in 1930 yielded numerous dinosaur footprints, most of which Sternberg was unable to quarry out. However, he made remarkably accurate plaster molds of the trackways, later reproduced in plaster or concrete for museum study.

Some researchers of the latter half of the twentieth century believe dinosaurs may have been warm-blooded, possibly more closely related to living birds than to reptiles. The discovery of dinosaur nests, some with eggs and eggshells and even fossilized embryos intact, helps create more complete images of the herds of dinosaurs—their procreation, eating habits, and migrations. From the study of the size and separation of the trackways, such as those found by Charlie Sternberg, paleontologists speculate on how these animals walked and ran, swinging their back feet forward to cover over the front footprints, so it appeared they walked on just two feet. Trackways also indicate that there were ancient stampedes and that dinosaurs probably moved at speeds up to forty miles (64 km) per hour and rarely let their tails touch the ground.

As early as 1918, Samuel Williston, a friendly competitor and contemporary of Charles H. Sternberg in their early hunting years, wrote in the *Transactions of the Academy of Science*: "The science of paleontology, the history of animal life upon the earth, has ceased to be merely the handmaiden of geology.... Lifeless fossils of half a century ago have become alive again, and their teachings have thrown a brilliant illumination upon the origin, relationships, taxonomy and genealogies of organisms.... This has united sciences of geology and biology and furnished the best proof of organic evolution."[9]

Charlie Sternberg was part of the bridge between the past and the future of the study of dinosaurs. Charles H. had taught his sons that finding, removing, and preparing fossils for exhibit—and then finding a market for them—was the whole purpose of the business. However, the father often fantasized about the period of the dinosaurs, letting his imagination carry him into a dream world where he watched great herds of dinosaurs at play. Charlie had no such fantasies. He simply wanted to know more about these ancient "monsters" and to be able to prove his theories.

When the Great Depression of the early 1930s slowed progress everywhere, the Canadian government felt no urgency to continue its dinosaur research programs: economic stability and progress took precedence. But, by 1935, officials decided that the many important specimens in the Steveville-Deadlodge Canyon area really ought to be accurately recorded. Consequently, the Geological Survey of Canada undertook a detailed topographical survey of the area with F. P. DuVernet in charge—and with Charlie Sternberg collaborating to locate and identify the sites of fossil discoveries.[10]

The work continued throughout 1936. Each site was located and then permanently marked by a brass plate set in an iron pipe with a concrete base. The project culminated in the publication of the *Steveville Sheet* (Map 969A, Geological Survey of Canada), giving an annotated list of all the finds, as compiled by Sternberg.[11] While preparing this material, Charlie did not lose his touch at finding

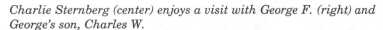

Charlie Sternberg (center) enjoys a visit with George F. (right) and George's son, Charles W.

Charlie's Steveville Sheet *listed Canadian fossil-find sites such as the deeply fluted, high, narrow pinnacles of Knife Ridge where two dinosaur skeletons were discovered.*

fossils, and was able to bring in a sizable collection of valuable items, including the skull of a new duck-billed dinosaur—the holotype of *Brachylophosaurus canadensis*—very similar to *Kritosaurus*.

In the summer of 1937 Sternberg explored the Manyberries and Comrey districts in southeastern Alberta. Russell had found well preserved vertebrate fossils in the Oldman formation, and Sternberg hoped to add to the list. He secured the skeleton of a large duck-billed dinosaur from east of Manyberries and found two skulls of a horned dinosaur, *Monoclonius lowei*, a type, according to Russell, previously known only from the Judith River formation of Montana.

Official work on vertebrate paleontology was again suspended during World War II; but as soon as restrictions were lifted, Sternberg was back in the field of the Red Deer River region.

Charlie spent the summer of 1947 in the "Upper Edmonton beds," northeast of Elnora at the north end of the Red Deer badlands on his last expedition for the Geological Survey. Sternberg and his assistant, T. P. Chamney, found three incomplete skeletons of *Triceratops* and *Tyrannosaurus* in the highest stratigraphic positions of any dinosaur remains known at that time. Here they also found a partial skeleton of a small, hornless ceratopsian *Leptoceratops*, originally made known by Barnum Brown. As Sternberg's excavation progressed, he found parts of a second skeleton, then a third. The last one had never been exposed to surface erosion, so included every bone—a truly remarkable find!

The next year the responsibility for vertebrate paleontology research and display was transferred from the Geological Survey to the National Museum of Natural Sciences, and a complete administrative separation of the two institutions followed in 1950.

In 1948 Charlie Sternberg received the rank of assistant biologist with the National Museum of Canada, and in 1949 he was elected a fellow of the Royal Society of Canada, the equivalent of the American National Academy of Science. He retired officially from government service in 1950, but continued to study and describe fossils in the collection of the National Museum of Canada until 1970.[12]

Charlie and his wife Myrtle had established permanent residence in Ottawa during their early years in Canada. They had three sons, all born during the busy years when Charles was spending much time more than two thousand miles (3,200 km) away. None of the three elected to pursue their father's profession, but all three earned doctoral degrees. Raymond, the oldest, wrote five articles for publication in scientific journals, all related to paleontology, but spent his active career teaching at the University of Victoria. Stanley worked for the Microbiology Research Council in Ottawa; and Glenn was head pathologist at Inland Hospital in Kamloops, British Columbia. At some point in their adult life, presumably during World War II, all three sons legally changed their names, taking their mother's maiden name of Martin to avoid the complications they believed the Jewish-sounding Sternberg would bring.

When the government of Alberta decided in 1956 to create a park in the Steveville Deadlodge Canyon badlands, Charlie Sternberg, at age seventy-one still considered an important resource person, was engaged as a consultant. Within a year he was able to see most of his ideas accepted. A park warden was selected, and in 1958 work on the project began. Sternberg was on hand and, as might be expected, found a few more specimens to add to the collection. Instead of removing them, however, the museum had a small house built over each of two incomplete skeletons, with windows to enable visitors to inspect the fossils where they were discovered. Later a nearly complete *Lambeosaurus* was found by the park warden and prepared for an outdoor display. This was Charlie Sternberg's last field project. He returned to the park in 1965, however, at age eighty, to supervise the collecting and display set-up of a nearly complete skeleton of another *Lambeosaurus*.[13]

The remains of approximately twenty-five different species of meat- and plant-eating dinosaurs have been found in the area of what is now the Dinosaur Provincial Park. Practically all of the area explored by the Sternbergs is included in the park area. It is one of the largest fossil museums in the world and an important tourist attraction.

One Sternberg fossil that went on display in 1969 at the Provincial Museum in Edmonton has an interesting history. It is an incomplete skeleton of the duck-billed *Lambeosaurus magnicristatus* collected by Charlie Sternberg in 1937, about seven miles (11 km) southeast of Manyberries in the extreme southeastern corner of Alberta. The skeleton was originally sent to the National Museum of Natural Sciences, but was released in 1966 on a permanent loan basis for the new Provincial Museum. At the time of release the skeleton was still in its original shipping boxes, with five blocks of bones, each wrapped in a plaster-burlap jacket and packed into straw-filled wooden boxes.

Sternberg had found the skeleton lying on its left side, with many parts badly eroded. One forefoot and the hind and forelimbs, both shoulder blades, ribs, neck vertebrae, and the skull were all well preserved, and there was a large section of skin impressions on the right side.

Sternberg had written in his field notes that the fossil had probably been exposed to the elements for more than a century and was not in good enough condition to make a free-standing mount. He wrote in his notebook: "This specimen is suitable only for a panel mount with right

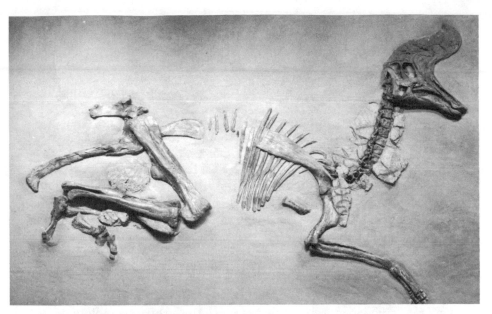

Lambeosaurus magnicristatus. —Tyrrell Museum of Palaeontology, Drumheller, Canada

side showing so I have put a layer of grey plaster on the weathered bones of the left side to give better backing for the mount. This plaster should hold the ribs, pelvic arch, and hind limbs in place. ... The neck and forelimbs are not so treated."

At the laboratory of the Provincial Museum, each plaster wrap was removed, and all bones painstakingly cleaned. Preservatives were applied to the bone as it was exposed, with special attention given to the skin impressions. In parts of the neck, the skin impressions were carefully left intact and made available for scientific study. A large area of the impressions was mounted with the skeleton, for an exhibit that gives the viewer a good idea of the skin of the duck-billed dinosaur.

As recorded in *Museum and Archives Notes*, No. 13, 1972, this *Lambeosaurus* was a large, crested hadrosaur that probably waded or swam—or possibly walked on its hind feet. The crest, plainly visible in the museum exhibit, is hollow and contained folded passages and chambers. Scientists believe these were sounding chambers, air storage chambers, and air traps to prevent water from entering the lungs when the dinosaur's head was submerged. Some researchers believe that membranes in these passages accounted for the dinosaur's sense of smell.

The *Lambeosaurus* was mounted on a large wooden framework with a piece of plywood placed over it. The skeleton sections were bolted into place with blocks of wood underneath for support. Styrofoam blocks held the forefoot and skull in position. Then the whole display was shellacked. This type of mount enables the viewer to see both bones and skin impressions, and allows the display to be moved easily. And moved it has been.

Along with other dinosaur material from the Provincial Museum, the *Lambeosaurus* now has a new home in the Tyrrell Museum in Drumheller, which opened in September 1985 and is named in honor of Joseph Burr Tyrrell, whose discovery of the an *Albertosaurus* skull in 1884 sparked international scientific interest that continues today. Operated by Alberta's department of Culture and Multiculturalism, the museum encompasses 11,200 square meters. According to a museum publication, Dr. Alan Charig of the British Museum of Natural History praised it thus: "There is nothing like it anywhere in the world. It will set new standards worldwide for what scientific facilities and public exhibition of dinosaurs ought to be."

The Sternbergs would have been overwhelmed and gratified to see this magnificent museum housing more than thirty complete dinosaurs, remains of flying reptiles and prehistoric mammals, marine invertebrates and sail-backed amphibians. Computers and audiovisual presentations span the gap of sixty-five to seventy million years as they help bring visitors an up-to-date visit to antiquity. Unfortunately, none of the four Sternbergs, whose explorations played such a vital role in the history of fossil hunting in Canada, lived to see the Tyrrell Museum; but their fossils there enlighten and delight more than a half of a million visitors yearly.

In his later years Charlie Sternberg retained an office in the Vertebrate Paleontology Laboratories of the National Museums of Canada. In 1960 the University of Alberta, Calgary, conferred on him the honorary doctor of law degree; and in 1974 Carleton University awarded him an honorary doctor of science degree. The honors came not only because of his many remarkable discoveries in the field and his scientific descriptions of these finds, but also because he had clearly bridged the gap between discoverer and scientist.[14]

He was listed for several years in *Who's Who in Canada*, was an honorary member of the Society of Vertebrate Paleontology, served a term as president of the Ottawa Field Naturalists Club, and was a member of the Professional Institute of Civil Service.

He found great satisfaction in several areas: he loved to find dinosaurs, and then to write descriptions of them and to speculate on their lifestyles. But he also loved to visit with children and explain these ancient animals to them. He welcomed visitors to his field camps and on weekends liked to conduct guided tours to explain recent discoveries. Perhaps he never ceased wanting to be a teacher, like his

Charlie Sternberg. —Royal Society of Canada

grandfather before him. He had a talent for storytelling and brightened his presentations with funny anecdotes. He was characterized by Russell, his good friend and fellow scientist, as a "steady, serious student, to whom the discovery and disclosure of dinosaur fossils was more a mission that a profession."[15]

In December 1977 Charlie, then past ninety-two, wrote a personal letter to a family friend about his life. It said, in part:

> I was born Sept. 18, 1885, married Sept. 18, 1911. Mother wanted me to be named Charles Mottram but I was baptized Charles Mortram. I saw the card, made out by the minister who baptized me in the church at Lawrence. I presume that he misunderstood father's writing, but as I was baptized Mortram, I suppose that is my name. We had the three sons and eleven grandchildren. ...

I was 92 last September and am in fairly good health. I try to walk two miles [3.2 km] each day and I curl [a game similar to shuffleboard] twice a week. I am still in touch with the Museum of Natural Science. I gave my library to the Museum. I still get some papers, especially on dinosaurs.[16]

Charles and Myrtle had been married sixty-five years before she died January 1, 1976, shortly before her ninety-first birthday. Charlie remained alert and was able to maintain an active interest in his career until very shortly before his death in Ottawa, September 4, 1981, a few days before his ninety-sixth birthday.

Russell summed up Charles M. thus: "A man who bridges the transition from the professional fossil collector who searched for unearthed prehistoric remains as a freelancer or an employee of an institution, to the academically trained paleontologist."[17]

CHAPTER FIFTEEN

New Horizons West

When Charles H. Sternberg left Canada, he went home, to Lawrence, where his laboratory-workshop was still intact. He finished his book about his Canadian experiences and was successful in raising money for its 1917 publication. He wrote several articles for magazines and managed to dispose of most of his property, but kept some storage space for fossils.

Then he and Anna made a major decision: they would move their headquarters permanently from Lawrence to the warmer climes of southern California, an area Charles was somewhat familiar with because he had made several trips to the Far West and had sold numerous specimens to museums at California universities.

Charles had tramped for fifty years over thousands of miles in the western half of Kansas, northern Texas, parts of Oklahoma, Wyoming, Washington, Oregon, Idaho, Utah, Montana, Colorado, the Dakotas, and Nebraska, as well as southern Canada. He had endured extreme heat and cold, hot searing winds, pelting sand, driving rain, and snow. Both he and Anna had made personal sacrifices and endured privations and separations. He had suffered the ague and an assortment of other ailments. Always he'd had to tolerate his stiff leg, deaf ear, and problems brought on by his pattern of rugged living. Now his dinosaur hunting days might be drawing to a close. But there were other worlds to conquer. Perhaps, he reasoned, all he needed to keep going was warmer winters.

He had written: "The weapons of the fossil hunter are pick, crowbar, chisel, soft brushes, cement, burlap, and plaster of Paris."[1] But the old hunter, now past seventy, was not ready to lay down his weapons, nor did he want to quit tramping over rocks and wastelands. There were areas, such as New Mexico and southern California, that he had not yet explored. It was time to move.

Sternberg was always proud of himself and his accomplishments. Neither modest nor humble, he always struggled for recognition and probably also suffered silently because he had such a limited education. This was compensated for, however, by his years of experience and the recognition accorded him. Such praise was comforting and reassuring. It was an incentive to push him a little farther to find just one more largest or "best of species" fossil.

Referring to his own experiences, and realizing he represented a fast-fading generation of explorers, he wrote:

> Sometimes when he [the fossil hunter] finds a skeleton projecting from the face of a cliff, he turns carpenter and builds a lofty platform on which to work as he removes it. Again, he calls in dynamite, and blasts his treasure unbroken from its natural sarcophagus of stones. There are not more than a half a dozen of us, but I doubt if any hunter, facing a charging elephant or an angered rhino, ever felt any greater thrill than we do when we stumble upon a new variety of monster, possibly the remote ancestor of these big-game animals of today.[2]

Early in the spring of 1921 Charles established headquarters in San Diego and found an assistant to help him start a new adventure. He claimed he just wanted to explore new country and study the fauna of the Cretaceous in New Mexico. But those who knew him realized he probably knew more than he was telling and had more specific plans than he cared to disclose.

Evidence of dinosaurs had been discovered in New Mexico long before Charles ever set foot in the state. David Baldwin, an assistant of Cope, had worked in 1881 northwest of the little town of Abiquiu, southwest of Taos, where many cliffs of Mesozoic strata rise and where many miles of late Triassic exposures are evident at the base of these cliffs. He found scattered bones that he had packed and marked "small and tender," and shipped to Cope, who recognized them as bones of a small, lightly built carnivorous dinosaur.[3]

Barnum Brown had been in the New Mexico territory in 1904 and found the skull of the duck-billed dinosaur *Kritosaurus*. Subsequently, United States Geological Survey explorers, operating in the same general area, made minor discoveries that C. W. Gilmore reported in 1919. Sternberg, an avid follower of anything Gilmore said or wrote, was inspired to do his own investigating.

The New Mexico expedition differed from all previous Charles Sternberg trips in its source of horsepower. For the first time he left the horses behind and drove a one-ton Ford truck. He and John Bender, his assistant, started from Los Angeles in May to drive eight hundred miles (1,280 km) through territory Charles had never before explored.

CROSS SECTION OF PETRIFIED TREE RICH WITH
AGATE CORRELIAN JASPR AND TOPAZ.
PETRIFIED FOREST, ARIZONA - NOTE THE DOVE

A postcard Sternberg bought on his trip to the Petrified Forest.

By the middle of the month he was in Arizona, roaming through the Petrified Forest Monument (later a national park), basking in a beauty he had never before seen. He was appalled at the deterioration of the fine old fossilized trees and the absence of either caretaker or custodian. Charles talked to anyone who would listen and would write in the 1932 revised edition of *Hunting Dinosaurs in the Badlands of Red Deer River*, making two suggestions that he felt would help preserve "this most magnificent of all the extinct forests in America." First he would have each great museum in the country, working with the federal government, remove a single perfect specimen and preserve it. Second, he would have a custodian or caretaker live in the forest and patrol it. He recommended heavy penalties for anyone who willfully disturbed or injured "any part of a petrified tree trunk."[4]

Charles and his party moved east into New Mexico during the third week of May 1921; but before he could launch any direct activity, he received word that his wife was seriously ill. He waited three days, until he heard that she would have to undergo major surgery, and then he headed home to California. Throughout all his years of roaming, this illness was the only incident that prompted him to write about his wife.[5] He was away from work for more than three weeks! In the second edition of *Hunting Dinosaurs* he wrote: "I waited to see her strength gradually renewed, and thanked God that He had spared me

Charles and Anna Sternberg.

that great sorrow of losing the faithful companion whose untiring help
through over forty years, has made it possible for me to carry on the
work of recovering, year after year, for the world's collections those
treasures beyond price, that lie buried in the Cemeteries of Creation."

George's wife, Mabel, took her children and headed for California
immediately upon hearing of Anna's illness. Mabel had long been a
close friend and in many ways a true daughter to her mother-in-law.
Since George was, as often, far away—this time in Canada—and
planning a two-year trip even farther afield, she headed west prepared to stay indefinitely. She was tired of being alone with her
children for long periods of time. Charles was pleased with his
daughter-in-law's arrival, saying "it was indeed a great comfort" to
have her with them.[6]

On June 14, Charles went back to New Mexico and was tramping
through ancient Indian ruins, fascinated by the enormity of the old
structures as he constantly sought answers to dozens of questions of
how and why. His interest lay not only in what fossils he might find,
but in the relics of all forms of life he saw.

Near Farmington in the northwest corner of the state, at the site of some Aztec ruins, Charles found Neil M. Judd, United States National Museum curator of American Archaeology, working with Navajos and four white men, excavating the ruins of the Pueblo Bonita (Chaco Canyon), said to be the most extensive in North America. Sternberg was overwhelmed and studied the formations extensively, trying to establish their characteristics and seeing himself as an authority on the geologic structures.[7]

When he described this trip in his revised book, he again fantasized as he visualized "uncounted monuments of every shape, castles with battlements upon their crests, balloon-shaped domes of Russia, pagodas of China, walls and minarets of Arabia, the mighty cathedrals of Europe, towers, pinnacles, needles, pyramids and obelisks of Persia, Siam and Egypt," adding that never in his life had he seen "such a glorious panorama of palace and hovel, of fluted columns and majestic buildings..."[8]

Returning to work after the three-week interruption caused by his wife's illness brought a revelation to Charles Sternberg: for more than fifteen years, he hadn't hunted fossils without at least one son to help him! Now, no longer a young man, here he was, thousands of miles from George, Charlie, and Levi, with all their experience, competence and understanding of his preferred methods of work. He was seventy-one years old and had an ailing wife. His work schedule seemed more strenuous than it used to. Furthermore, he was about to hire two Indian boys as helpers!

All of his life he had been taught to be wary of Indians. He remembered the Mohawks in New York State when he was a child; the frequent incidents and alerts in the Kansas of his youth; the encounters when he was with Cope in Wyoming and Montana in 1876; and his brushes with the Bannock War in Oregon. He also remembered those times, a few years later, when his wife, a young mother with little children to care for, had raised and transported produce to feed Indians at the Haskell Institute in Lawrence. Times were changing even then. Now here he was, about to employ young Navajos to replace his sons as working assistants.

Under contract to Dr. Carl Wiman to work two summers for the University of Sweden in Upsala, Sternberg hired Dan Padilla and his team of Navajo ponies, and a few weeks later hired Ned Shouver, another young Navajo. Both boys worked for him all summer. They were much better at washing dishes than cooking, he discovered, but they were dedicated to their labor in hunting fossils. Since there were few other people in the entire area, Charles soon learned to lean heavily on the boys for loyal support. His half century of fossil hunting had taught him that it was futile to explore while riding—on anything—so he turned the truck over to his assistants, and again tramped on foot.

Charles quickly molded the Navajos into excellent fossil hunters. Misunderstanding their employer's motives, the Indians nicknamed Sternberg "Gold Hunter," a name he found abhorrent, deeming his treasures more rare than gold or silver. His resentment of the name only caused him to work harder. Finally, late in the season his efforts paid off when he found a complete example of the horned dinosaur *Pentaceratops*.[9]

Collecting this dinosaur occupied the team for the rest of the summer, requiring weeks of strenuous labor laced with frustrations. Charles tried diligently to maintain his tried-and-true schedule, starting his day as soon as dawn pierced the darkness. While his helpers brought in the horses, left hobbled for the night on a nearby mesa, Charles prepared breakfast—an act of self-preservation, he insisted. Leaving the boys to clean up after the meal, he left camp on foot to reach the fossil beds in the Kirtland shales. He traveled along a valley with cliffs and buttes on either side. As he walked, he noted every fragment of bone he found projecting from the cliff or slope; and if it looked at all promising, he carefully marked it with a pile of stone, to make identification easier for a return visit. Loose clay along the slopes crumbled easily as he walked and occasionally sent him sliding down into a gorge below—unless he had a pick handy to use as an anchor for his footsteps. Every time he deliberately ventured down a slope he had to make sure of a good place to land in case he tumbled. At first he deliberately sat and slid down the slopes, but soon learned to stand erect and use his pick as a cane for balance and support.

Sudden downpours made the heavy clay soil so adhesive that it clung by the pound to his boots and exhausted him. But despite such unexpected showers, Charles maintained: "...the successful fossil hunter goes carefully over every square foot of exposed surface, the mind wholly occupied with this one thought, and the eyes seeing nothing but the object of the hunt. The most valuable specimen may be easily passed over, especially if the mind dwells on something else."[10]

In helping Charles remove the *Pentaceratops*, the boys laid bare a floor eighteen feet (5.4 m) long, ten feet (3 m) deep, and with a wall six feet (1.8 m) high next to the bluff. But just when he thought he had it all, he accidently stepped on a hind foot and destroyed it. Soaking strips of burlap in plaster, he began wrapping sections as fast as they were brought out into the open air. Days were getting short, and the nights were cold. The plaster was used up, so he sent Ned to Farmington for more. The store had none on hand, but promised to get some "immediately." Ned didn't wait, but returned empty-handed after a four-day journey to find a disgruntled boss. Trying to be patient and to make himself understood, he sent both boys back to the store; and five days later they again began taking up the sections and wrapping them. The weather worsened. Wet plaster and cold temperatures cracked the skin of Sternberg's thumb and finger, drawing blood. His only home remedy at camp was a can of shortening, which he rubbed generously into his fingers as he sat around the warm campfire. Repeated applications eased the pain, and work continued.

Sternberg feared a sudden snowstorm and realized that when it was time to go home he would have to travel eight hundred miles (1,280 km) in a truck that did no more than fifteen miles (24 km) an hour—and first would have to get the *Pentaceratops* to Farmington. Time was precious, and he always had to contend with the language barrier between him and his young assistants. Sometimes it was almost impossible for him to make them understand the urgency of the work.

Finally, all sections were wrapped and ready to load. Charles used a tripod, pulley, and ropes to lift the fossil blocks into the wagons. The tripod, constructed from fourteen-foot (4.2-meter) long four-by-four (10-by-10 cm) timbers, was meant to operate with one boy stationed at one post and the other at a second. As Sternberg started the lifting process, the tripod slipped and collapsed. The boys narrowly escaped the crash and lay on the ground laughing, not worried about the fossil or their own possible fate.

Each young helper had a wagon: Dan, a four-horse outfit; and Ned, a two-horse one. Eventually, both wagons were loaded, and the boys were ready for their trip to Farmington with the *Pentaceratops*.

Charles, confident that he knew the route the boys would take with their wagons and that he could walk across the valley of Meyer's Creek and meet them on the other side on level ground, tried to explain to the boys what he had in mind. They all set out. Charles followed a road he assumed they would take, but somehow they missed connections. The boys had planned to stop at the Davis Store about thirty miles (48 km) northeast of where they started; but, when night came, Sternberg was totally lost. He broke off dead branches from a cedar tree for a fire and found leaves for a bed, but had no covers, only the clothes on his back. He started walking again at dawn, sure that he would soon reach the store and find warmth and food. But again he was disappointed: he was in a strange place with nothing familiar, and no store in sight. He began backtracking, fearing he would have to retrace the entire trip; but he came upon an Indian hogan and at length persuaded its owner, for two dollars, to take him to the Davis Store. It was only about a mile (1.6 km) away, but that might as well have been a hundred for all Sternberg knew. He concluded his two dollars was money well spent.[11]

185

As Charles prepared to discharge his Navajo assistants from his employ, his thoughts raced back over the summer's adventure. It hadn't been so bad! In fact, it was pretty good! He had learned to appreciate the boys' loyalty and skills, as well as their good humor. They had learned how to find a fossil and excavate it. They also had learned to put business before pleasure. The camp had been only about eight miles (13 km) from a store at Kimbeta—a settlement that, like many others Sternberg identified in his accounts, no longer exists. Near the store lived a widow with "a hogan full of pretty Navajo girls," whose presence made going to the store for supplies a special treat for the boys. Charles also recalled the time when both boys had wanted to attend an Indian celebration, a dance and horse race— something no Navajo felt he dared miss. But both boys sacrificed their own pleasure in order to make a trip to Farmington for supplies Sternberg needed.

He remembered their pleasure at finding fossil turtles: together they had collected sixty, representing five families and many species. Their finds included the rare *Aspiderites*, with its round carapace or shield; and the *Adocus*, which had a smooth carapace and was twice as long as it was wide. Altogether they had taken out fourteen hundred pounds (630 kg) of Cretaceous vertebrate material that summer.

So it was with a good feeling that Charles paid the young Navajos their wages after they delivered their last load to Farmington for shipment east. He left New Mexico on November 27, 1921, glad to return to Anna in California. It was indeed time for Thanksgiving, he realized.

Levi Sternberg spent a year in 1923-24 with his parents in California—one of the only times in his life he had an opportunity to work alone with his father. Twenty-eight years old at the time, he requested a leave of absence from his work with the Royal Ontario Museum and apparently enjoyed a relatively quiet year free-lancing with his father. The only specimen Charles mentioned in his records of this period was a *Parasaurolophus* that Levi found. Giving no other details about how big it was or where it was found, he described it as having a narrow crest that extended back of the head twice the distance of the length of the skull. Levi sent it to Toronto University.

Sternberg's last years of fossil hunting were done a little closer to home. He explored the Pleistocene fossil beds at McKittrick, California, west of Bakersfield, at the south end of the San Joaquin Valley. A house on an old oil lease owned by Midway Petroleum Company was made available, and the Sternbergs lived there from 1925 until 1927. Charles set up a workshop in half of the structure, and for the first

Charles and Anna's house in McKittrick.

time in all his years of preparing fossils, he could do his laboratory work almost on the spot of discovery.

He found a vast cemetery of bones of birds and mammals, including a bison, much larger than the Recent buffalo. With horn cores measuring forty-four inches (112 cm) from tip to tip, the bison was eleven feet (3.3 m) long from end of chin to base of tail. Sternberg recorded that he discovered horses and camels, llamas, antelope, bison, saber-toothed tigers, wolves, lions, and bears, all lying together—the first such collection he had ever found.[12]

Sometime earlier, the University of California had removed a large mastodon skull and many other bones from the same quarry. Many of the bones Sternberg found were asphalt-covered and required immediate attention to be cleaned, then soaked in diluted shellac to prevent disintegration. He mended broken bones with cement of gum arabic and plaster of Paris. He had been told that this quarry would resemble the La Brea tar pits where hundreds of mammals—tigers, wolves, horses, and other species—were trapped in ancient ages, but Charles determined that in the McKittrick Quarry the bones had been carried in by water, not trapped as in the La Brea pits. Here he found the asphalt extending downward only about two feet (61 cm).

His efforts were rewarded with about twenty skeletons, but equally important to him was the fact that his explorations attracted the attention and approval of vertebrate paleontologists at the University of California-Berkeley, and at California Institute of Technology, for which he was working at the time. He was asked to continue his excavations, working at the nearby University Quarry.

There he found a pocket ten feet (3 m) below the surface filled with remains of Recent animals. Since he had assumed that all the animals he would find would be of the Pleistocene period, he had to adjust to accept the fact that here were examples that bridged the ages. But the

The McKittrick oil field yielded many fossils for Charles Sternberg.

old fossil hunter was tired and at long last willing to let someone else solve the mysteries of how the later bones happened to be there.

Nevertheless, when he received word in 1928 that he had been granted permission to explore Baja California, near Catarina, to look for ammonites, the temptation was too great. With one assistant—Dr. Peter Alpeter, a Mexican of Bonita, California, who supplied a pack mule—Charles drove south from San Diego over the San Miguel Mountains. They stopped at an onyx bed, and in sandy concretions along a narrow gulch, in level land near the ocean, they found huge ammonites. They secured thirty-five specimens that measured from seventeen to twenty inches (43 to 51 cm) in diameter and often weighed a hundred pounds (450 kg).[13]

Alpeter hauled them to the car, parked a mile (1.6 km) away, then to the wharf where they were packed in metal drums and shipped to San Diego. Eventually these specimens found their way to museums all over the world.

This was Charles Sternberg's last field expedition. He summarized his experiences since leaving Canada by writing a five-chapter addendum for the 1932 edition of *Hunting Dinosaurs in the Badlands of the Red Deer River*. He and Anna continued to live in southern California until her death in 1938. Now eighty-eight years old and truly alone, Charles realized it was time to give up his independent living and make his home with his sons. He had traversed the United States many times. He had shared in the transition from horse and wagon to air travel. He had long before intoned the twofold purpose of his life. The first, the desire to add to human knowledge, was, he said, his life's greatest motive. The second, the hunting instinct deeply implanted in his heart, he described as follows in *Hunting Dinosaurs in the Badlands of the Red Deer River*, Alberta, Canada:

> Not the desire to destroy life, but to see it. ... The man whose love for wild animals is most deeply developed is not he who ruthlessly takes their lives, but he who follows them with the camera, studies them with loving sympathy and pictures them in their various haunts. I love creatures of other ages and want to become acquainted with them in their natural environments. They are never dead to me. My imagination breathes life into the "valley of dry bones."... No matter what the common herd may say about me, I have done my humble part toward building up the great science of paleontology. I shall perish, but my fossils will last as long as the museums that have secured them.[14]

Although Charles Sternberg's reputation depended largely upon fossil mammals, he also contributed heavily to invertebrate paleontology and the study of fossil plants. His first discoveries, impressions of ancient sassafras leaves sent to the Smithsonian, prompted him to dedicate his life to fossil hunting. Some of his final discoveries were of invertebrates. He was among the first to collect entire individuals of the very large bivalve *Haploscapha*, the stemless crinoid *Uintacrinus*, and the ammonite *Pachydiscus*.

In addition to his two full-length books about his own experiences, he wrote at least forty-five articles on various geologic and paleontologic subjects, and contributed to at least eleven volumes of the *Kansas Academy of Science Transactions*.

When he left California in 1939, he visited George, who had settled in Kansas, then went to Canada, dividing his time between his two sons there. He finally settled in Toronto with Levi, and died peacefully in his sleep on July 21, 1943, at age ninety-three.

It was not the general public who paid tribute to Charles H. Sternberg during his lifetime, but his peers and scientists he had known throughout his wide span of years. Dr. William King Gregory of the American Museum in New York once said to Charles, "Your whole life and work have placed all paleontologists under lasting obligations to you."

Douglas Preston, at the time with the American Museum, wrote: "Science requires the activities of two very different kinds of people— the brilliant thinkers and synthesizers, and the hardworking but unimaginative compilers of data. ... Sternberg was a member of the latter group. ... He never served on the staff of an institution and never claimed to be more than a fossil collector, although he did publish descriptive papers on his finds."[15]

The late Dr. Charles Gilmore, curator of Fossil Reptiles at the Smithsonian, once called Charles Sternberg the only American collector of his time whose livelihood was largely gained from the collection and sale of fossil specimens. Such a career called for many privations and personal sacrifices, and the remuneration was often inadequate and discouraging; but his enthusiasm for the work never deserted him, according to Gilmore.[16] Certainly the Sternbergs were the first family group ever to hunt fossils both as a way of life and means of livelihood.

David A. E. Spalding of Alberta's Provincial Museum in Edmonton wrote in his introduction to 1985's third edition of *Hunting Dinosaurs in the Badlands of the Red Deer River* that Sternberg was sometimes garrulous but never a bore. He was not always up-to-date in areas of science outside his specialty, yet was willing to change his fondly held ideas when presented with new evidence. Spalding praised Sternberg, describing him as scientific and imaginative, objective and enthusiastic. He pioneered scientific techniques and made many of the most important fossil discoveries of his time. Wrote Spalding: "We can

Charles H. Sternberg. —Kansas State Historical Society

experience the taste of blowing sand, feel the hot sun on our backs, and buckle down to the back-breaking but exciting task of revealing for the first time, some unknown monster of former ages." Spalding also pointed out Sternberg's deep-rooted religious convictions, suggesting that Sternberg felt he was doing God's work and that by writing about his discoveries he was promoting his beliefs and encouraging his readers to wonder at the "creatures of the misty past ... these lords of creation."[17]

Charles Sternberg often wrote of himself. "My own body will crumble in dust, my soul return to the God who gave it, but the works of His hands, those animals of other days, will give joy and pleasure to generations yet unborn."

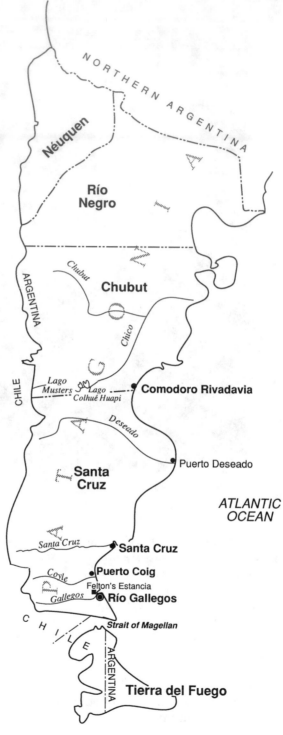

SOUTH
AMERICA

Patagonia

NORTHERN ARGENTINA

Néuquen

Río
Negro

Chubut

Chubut

Chico

ARGENTINA

CHILE

Lago
Musters

Lago
Colhué Huapi

Comodoro Rivadavia

Deseado

Santa
Cruz

Puerto Deseado

ATLANTIC
OCEAN

Santa Cruz

Santa Cruz

Coyle

Puerto Coig

Felton's Estancia

Gallegos

Río Gallegos

CHILE

Strait of Magellan

ARGENTINA

Tierra del Fuego

CHAPTER SIXTEEN

The Ends of the Earth

I f George Sternberg had picked up a map and put his finger on a spot that would place him at the point on the earth opposite his 1922 Canadian field of operations, he quite possibly would have touched the destination of his next fossil hunt—the Patagonia region of southern Argentina, dubbed the "uttermost part of the earth" by author and explorer E. Lucas Bridges. In Patagonia George would be as close to the South Pole as he had been to the North Pole while in Canada.[1]

Considered unapproachable for ages, the southernmost nine hundred miles (1,440 km) of Argentina attracted the scientific world in 1833 when a British ship sailed through the area, doing geodetic surveys. On board was Charles Darwin, who returned to England with an array of fossil specimens entirely new to science. From his discoveries, scientists realized that the fauna of Patagonia was unlike any in the rest of the known world, that this region had been geologically isolated for perhaps sixty-five million years of earth's history.[2]

In the 1880s two Argentinean brothers, Carlos and Florentino Ameghine, explored the region and published reports of their discoveries. Paleontologists from Princeton University and the American Museum of Natural History, as well as Andre Tournouer of France and Handle T. Martin, sent by the University of Kansas, also began collecting in the region. The area explored included the states of Neuquén, Río Negro, Chubut, and Santa Cruz; and it was in this small coastal area at the southern tip of the continent that dinosaurs were first discovered in South America.

In 1882 Santiago Roth hunted dinosaurs in the vicinity of the city of Neuquén and found bones, which he sent to England. About 1910

Dr. R. Wichman made an expedition and developed quarries in the southern part of Río Negro where he found remains of armored dinosaurs, as well as bones of swamp-dwelling sauropod dinosaurs. His 1916 paper about his discoveries attracted other hunters. His efforts, the first large-scale digging for dinosaurs in Argentina, produced the most important collections that had been taken out prior to 1916.[3]

The Museum of La Plata in Buenos Aires, a leader in the study of Argentinean fossils, launched an extensive dinosaur hunt about 1918, establishing a base east of Neuquén. Roth headed explorations farther south in 1921 and 1922 and found more Cretaceous dinosaurs, including bones of gigantic sauropods, which added to the holdings of the La Plata Museum.[4]

News of these discoveries had spread throughout the scientific world and reached Barnum Brown and his good friend Elmer Riggs in Wyoming. Inspired by their teacher, Samuel Williston, the two young men had planned someday to make their great adventure to the southern tip of South America. There they would find rare fossils and indeed startle the world! They were only beginners in fossil hunting at the time, but they never lost sight of their dream. For Riggs, the chance came when he was chosen to head several expeditions to South America for Chicago's Field Museum with the generous financial support of Marshall Field.[5]

Field Museum group in Canada, before leaving for Patagonia. E.S. Riggs, head of the South American expedition, is in center; J. B. Abbott is at the far right. George Sternberg took the picture.

George bids an affectionate good-bye to his faithful team, just before his departure on the South American expedition to Patagonia.

But George Sternberg, not Brown, was chosen to accompany Riggs on his first great adventure under the constellation of the Southern Cross. The Field Museum planned expeditions covering a period of five years to study the geology of Argentina and its adjacent states and to collect fossil mammals. No North American institution had undertaken such a venture since the Princeton expeditions of 1896-99.[6]

Riggs, Sternberg, and John B. Abbott, all employed by the Field Museum staff, were working in Alberta, Canada, at the time. When the authorization for the expedition to Argentina was confirmed, the three wasted no time closing their affairs in the north land and getting to Chicago for final briefings.

Sternberg men were noted for making their own decisions and expecting their wives to be receptive and cooperative. But when George told Mabel about his wonderful opportunity to spend two years at the other end of the world, not only was she unenthusiastic, but she was just plain disagreeable. She was in California at the time, fulfilling a daughter's role in the life of her sick mother-in-law. Her children were pre-teens, and she was weary of the increased burdens she already bore because of George's continued long absences from home. He listened to her objections, but made his own decision, agreeing to go to Patagonia.

The party embarked on November 15, 1922, for a seventeen-day voyage to Buenos Aires, and, after a nineteen-day delay, went on south to Río Gallegos on a South American vessel that plied the coast of Patagonia, carrying mail and supplies and taking back raw materials. There were no north-south railroads and only a few short ones running west to the mountains and Chile.[7]

Almost immediately after arrival, Riggs learned of a new law that protected Argentinean institutions but caused trouble and delay for foreigners. The law mandated that all collected material be inspected by Argentine officials before it could be shipped out of the country. Any new specimens could be confiscated. Half of any series of desirable specimens could be withheld, and field work could be prohibited in certain areas. It took Riggs almost a month to work his way through the restrictions and red tape and to convince the local authorities the expedition was honorable and deserving of their cooperation.[8]

The directors and staffs of the Museum of La Plata and the National Museum of Buenos Aires extended many courtesies and supplied aid in a number of ways to the expedition. This made it possible for Riggs and his men to begin work almost immediately, although an undercurrent of recent trouble prompted Riggs to write to the director of the Field Museum: "It is said there was open rebellion in this territory a year ago and some 400 men were executed. ... Please do not communicate this to our families."[9]

Riggs followed the route used by Princeton paleontologist John Bell Hatcher in 1896 and established camp on Felton's Estancia, a sheep ranch on the north bank of the Río Gallegos, about forty miles (64 km) from the city of Río Gallegos. There, at the eastern end of the Strait of Magellan in the Santa Cruz province, the icy air swoops off the

George labeled this: "Hill Station, Río Gallegos."

196

One of George's postcards from Patagonia shows one of the better roads winding up toward the pampas.

Andes, and the endless windswept pampas challenge the ranchers and even nature herself. The men worked the north bank of the river in the Santa Cruz formation, which dates to the early Miocene or Oligocene epoch, about twenty million years ago.[10]

The formation, cliffs about three hundred feet (90 m) thick on the eastern coast of Argentina, lies near the Straits of Magellan, west of the Falkland Islands. The hunters found fossils at all levels, but the cliffs were steep and inaccessible to people on foot. Tides rose to a maximum of fifty-eight feet (17.4 m), dissolved, and washed back, exposing as much as two miles (3.2 km) of the ocean's floor and often dislodging huge stone blocks that slowly disintegrated, leaving vast quantities of clay, gravel, and rock. The softer material was constantly washed as in a great natural sorting pan, leaving behind fragments of fossils. The formation eroded by the tide revealed skeletons of fossil vertebrates mostly different from those found in North America.

Riggs reported about the fossils to the Field Museum: "We find them appearing like the broken ends of brown and hollow rootlets exposed at the surface of the fallen blocks of stone, accumulated at the shore. We find them exposed in the ledges of sandstone and in the layers of clays on the tide flats where twice daily they are swept clean by the advancing and retreating tides."[11]

197

An element of mystery always surrounded new territory, and almost anything could be expected. When a man introduced himself to Riggs as J. G. Wolfe and said he was a museum curator in Río Gallegos, Riggs listened to an incredible report of a "Tertiary human skull" and an "enchanted city." Then he employed Wolfe to take the party to the site. They set out on a 500-mile (800-kilometer) journey north to El Paso de Santa Cruz, where the skull was supposedly discovered, and learned that it had attracted attention of an English nurse working in the area in 1916. She had possession of the object and had proudly showed it off to Wolfe. Investigation revealed that the woman was mentally unsound and that the 22-pound (10-kilogram) "trophy" to which she clung was only a very unusual stone, with a head-like shape. The "enchanted city" was another disappointment, as it was really only an intrusive bed of lava, or dike, that had filled a fissure in the clays that surrounded it, then eroded, creating unusual shapes. Local residents saw nothing noteworthy about it. The "city" was a figment of the imagination.[12]

After Wolfe led the party on other short, uneventful expeditions, Riggs concluded that the man had no scientific training—only wanderlust and a fine imagination. Riggs dismissed him, and the company set about the business that had brought them to this end of the earth.

Roadways in the area generally ran parallel to the shore, but were several miles inland and high above the water line. Occasionally, about every ten to twelve miles (16 to 19 km), valleys or small streams

George Sternberg poses beside the expedition's light truck.

Having been lowered down the cliff, the roadster serves as packhorse.

offered a passageway from the road above down to the seashore. Sometimes the fossil hunters could travel on horseback, but in many cases they could proceed only on foot. Once, down at sea level, they found a safe passageway several miles long, suitable for horses—or a light car or truck, if ever they found a way to get a vehicle down to that level.[13]

Getting fossils from the point of discovery on the shore or out on the tide flat up to the roadway and the base camp offered challenges daily. Generally, the men deposited each day's specimens together in one cache above the expected range of the incoming tide; then, whenever possible, they took them on saddle horses to the top of the cliff. One time the men used parts of a broken rowboat to contrive a sled that they drew along the shore for several miles with a horse pulling a long tug rope. But usually the specimens were carefully bundled and secured in saddle bags on either side of a packhorse for the trip up the cliff.[14]

The expedition's equipment included a light truck and a small roadster. Working on the south side of Río Coyle, the men found that the stream valley approached the beach without excessive obstruction; using rope and tackle, they lowered the little car down the steep cliff so they could drive it on the hard beach sand below. This done, they were able to speed along for about ten miles (16 km), pick up a 400-pound (180-kilogram) load of specimens, return to a base safely above tide level, deposit the load in the cache, and return. The little roadster made as many as four or five trips between tides.

Life revolved around the tide. The struggle was always to find the fossils, get them out, and get themselves and their equipment out before the tide came back in. Other complicating factors were the sudden winds and rain that occurred frequently.

Camp organization was kept simple. Abbott cooked breakfast and packed lunch. George took responsibility for repairs and numerous

The notation on the back of this postcard from one of George's photo albums reads: "High tide along the beach just in front of Rivadavia— Patagonian beds. Here's where we found fossils."

maintenance jobs. Each man cleaned, shellacked, and wrapped his own fossils and kept the numerical records of his discoveries. Riggs brought supplies when he made trips into the nearest marketplace. Abbott and Sternberg started out working as a team; but after a while Abbott complained that he was tired of following after Sternberg and thought it would be better if they went in opposite directions once they reached the tide flats. Sternberg's note of reply was that he didn't care how they worked, just as long as Abbott was happy.

If the morning brought stormy weather, the trio did odd jobs around camp. The need for wood was a constant source of frustration as Sternberg's numerous diary entries show: "Abbott and I spent an hour and a half searching for wood and came back without, only a small armful," he wrote. But sometimes on little trips looking for wood he would find a part of a skeleton, a skull or other object of interest: "I found an old sea lion and saved the skull, also a small, young seal, and saved the skull." He boiled these until the meat fell off; but most of the teeth fell out of the sea lion skull, and the skull of the seal fell apart. He still salvaged the pieces and laid them out to dry.

Riggs brought three horses from Hollidays, a supply center about thirty-five miles (56 km) from camp, and Sternberg claimed a small one named Niggar, who became a faithful and loyal companion. Sometimes George rode him; sometimes he was a packhorse; sometimes he towed a heavily laden sled. Work always began early in camp. Rounding up horses often proved to be more than a one-person job, and Sternberg assisted Riggs. As quickly as possible, the supplies for the day were loaded on the horses, and the treacherous trip down the cliff began.

George's caption reads only: "We welcomed opportunity to buy supplies here."

At low tide the party could travel for several miles in either direction, sometimes as much as five to ten miles (8 to 16 km) away from the shore-based temporary camp. Always the three men felt the need to stay in touch with each other and to hurry back to the designated point as soon as the tide came in. Sternberg described a typical day:

> March 13, 1912. It was 10:30 by the time we got to work. What a waste of time when we might be camped very much closer than here. We found our horses, which Mr. R. had us hobble, had wandered a good ways down the slide and we had to rush to get out before the water cut us off. We put our horses into tide surf and struck some bad holes and soft sand but we got by and reached camp at 6:15. I brought in a post, dragged it from the cinch of my Niggar horse. He is a good horse to ride.
>
> I found an *Astrapotherium* skull and though the right arch is weathered away, the rest is present. I hope to take it up tomorrow.
>
> I also got a fine skull of a rodent with one lower jaw present.[15]

The next day, although the crew was up and busy by 6:30, the tide was so high that they could not get down to the work place until almost

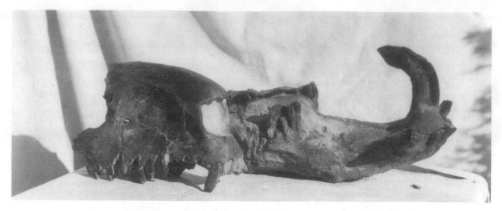
Astrapotherium *skull from Cape Fairweather, Río Gallegos.*

noon. George got busy immediately on the *Astrapotherium* skull and found that the lower jaws were present, but out of place and slightly crushed together. The tusk, which was exposed, measured seven inches (18 cm) beyond the jaw or upper tusk, while the lower jaw was not so long. By the time he got the skull cleaned up, George realized he had found a prize specimen.

One rainy morning Riggs took the rifle with him when he went for the horses, saying he would bring back a rhea for supper. He didn't find one that morning, but brought in a load of mushrooms. When the rain subsided, Riggs went over the pampas to the ranch at La Costa for food and supplies and down on the tide flat from there. He did not return until 9 p.m., and reported his horse had gotten away. He found no one at the La Costa ranch, so his long trip was in vain.

Winds were violent, but seldom kept the trio from accomplishing something. George wrote of one day in particular when it was too windy to go far. He contented himself by wrapping and labeling fossils:

> We rode out across the pampas to the water hold gap where we packed up all the fossils we had stored there. I had 11 numbers [each specimen was wrapped and assigned a number] and I plastered six of them. The plaster sets too fast, and I had a job getting the rags on before it set on me.... I collected my *Glyptodon*.[16] It proved to be all in sight except the tail club.

Often it was impossible to know just what sort of fossil the fragments had been. George reported finding a good part of a skeleton of a small individual with a very long skull. He had hard luck and, in handling the specimen, broke the skull into several pieces with his thumb. He had two or three feet (61 or 91 cm) of vertebrae and ribs, but was never sure just what he had found. He saved the pieces, wrapped them in rags and plaster, and shipped them back to the museum anyway.

Riggs left early one day on a trip to Puerto Coig, about thirty-five miles (56 km) by the route he took, although it was only two or three miles (3.2 or 4.8 km) across the bay from camp. He was gone until night, but returned with a case of gasoline, milk, and baking powder—all very essential items. Because they were dependent upon their own baking, there was no time to wait for yeast dough to rise, and quick bread with baking powder was a daily item.

The next day Abbott and Sternberg rode over the pampas to a new place where they could get down to the coast, just above the cache that held their collection of previous days. They saw herds of rheas, llamas, guanacos, upland or Magellanic geese, and buff-necked ibis, as well as black-chested buzzard eagles and Andean condors. They spent most of the day plastering and wrapping their specimens so they could be taken up to camp safely. George had fifteen in that lot and realized that he had numbered one hundred specimens in all. It was March 19, and they had been hunting only since January.

When the two reached camp that night, Riggs triumphantly showed them a dressed rhea, ready for the next day's evening meal. George said the meat was good, although coarse and wild-tasting. Supplies were running short. They were out of oil and candles, and so improvised by using a container of mutton tallow with a rag in it for evening light. Flour, baking powder, and fresh meat were gone. Heavy showers kept Riggs from going for supplies, and no dry wood was available. But they survived and kept working.

One day they made an unexpected discovery—a large quantity of oyster shells. The men took up more than thirty pounds (13.5 kg) and often found shells with both halves intact. "They are very good specimens and there seems to be only one species," Sternberg noted. He also found most of a skull of a sloth.[17]

A fourth member of the party, identified by Sternberg only as Miller, joined the group in late March, just before they were preparing to move camp. Abbott, Miller, and Sternberg took Miller's car down the cliff and brought in five loads of fossils, three from "the lower places" where they had piled up their finds and "two from the water hold." Sternberg recorded: "We had to all push to get the car back out of the water hold and got badly splashed as there are no fenders on the car." The entire collection was taken up to camp in two truckloads, but some material shattered into fragments in the hard rain and the rough ride.

Life was never easy, and ingenuity was taxed to the limit day after day. Usually Sternberg and company used grass as packing material for fossils, but in the wet season finding a substitute became hard work. They had to pull up grass, roots and all, then clean off the mud: the ground was too soft to hold the roots when they tried to cut off the tops of the grass.

After waiting several days for a change in weather, on March 29 Riggs pronounced the road sufficiently dry to be passable, and they began to pack for a move. They put together a light outfit for a temporary camp near the La Costa dipping pens, some six miles (9.6 km) south of La Costa. Three extra men, replacing a roof that had blown off the shearing shed, showed up for dinner. No one turned away a visitor at mealtime. The food supply was practically exhausted, but they ate what was available and then headed for La Costa. There the cook let them have one-fourth of a mutton, some flour, and one teaspoon of baking powder, with the promise of more soon, when the *capataz*, or foreman, returned with the key to the storehouse. They also borrowed oil and kerosene.

The roads were still so muddy that the fossil hunters could make no progress until they wrapped rope around the wheels of their vehicles. They reached their destination about six in the evening, made camp in a sheltered spot near the water with a mess tent, stove box, table, and two cots. The precious teaspoon of baking powder went into biscuits for breakfast.

The next day started on an optimistic note. Sternberg had a productive day, finding two small skulls of *Adinotherium*, a small Miocene *Toxodon* built like a short-legged rhinoceros. The first *Adinotherium* was not good, but the second had incisors all the same length and appeared to be a good specimen. The tide chased George out before he got the lower jaws collected, but he took note of the location so he could return the next day. On the way back to camp he loaded up with wood and found himself farther from base than he thought. When he finally arrived, at 7 p.m., two men from La Costa were there eating. Fortunately they had brought supplies, including baking powder and milk.

Working alone on a lovely day, Sternberg found an armadillo skeleton, with lower jaw and vertebrae, limb bones, and ribs. He could not see the skull but believed he would find it inside the mass he was taking out. He also found the lower jaws of the *Adinotherium* he had taken out earlier. He walked several miles on the flat, seeing hundreds of starfish and forty or more large crabs, and he picked up a collection of unusually pretty shells.

As he turned the shells over and over in his hands, his thoughts turned to home. It had been more than four months since he left his family, and it would be a lot longer before he would return. He realized he was homesick. The shells were for Ethel, his eleven-year-old daughter. His wife, Mable, hadn't wanted him to go on this expedition

George Sternberg's son, Charles W., and daughter, Ethel, lived in California with their mother while George was in Patagonia.

and might not be waiting when he returned; mail from home had been very limited and not reassuring. They had been separated too many times for long periods all the years of their marriage. But little Ethel surely would appreciate the beauty of these shells and the knowledge that her father had thought of her when he was so very far from home. Tenderly he wrapped the shells and laid them aside to be packaged carefully and separately for their long journey north.

George left the shore at 6 p.m., and it was after 8 o'clock when he reached camp. He must have walked at least five miles (8 km), he concluded. Abbott had eaten and cleared away the food. Sternberg ate a cold supper, put away his little package of shells, and sorted through the day's discoveries, which included a couple of pieces of whalebone from a sperm whale carcass he had found on the way to camp that night. Riggs had reported seeing the same carcass earlier.

A few days later George packed his gear to go down to an old camp to take up a *Prototherium* specimen that the tide had prevented him from securing earlier. He rode Princess, the sorrel, leading Niggar as a packhorse and setting out about noon across the pampas on the twelve-to-fourteen mile (19-to-22 kilometer) ride. The level land was very muddy, and water filled all the lagoons. He enjoyed the wildlife— herds of guanaco, flocks of wild geese, hundreds of flamingos and almost as many rheas. George shuddered a bit as he remembered the evening when Riggs had killed and cooked the rhea. The wildlife was so beautiful; he preferred it alive and free in its natural habitat.

George took a wrong turn and found himself west of the ranch buildings when a heavy shower came up, and he had to go about eleven miles (18 km) farther than he had anticipated. The horses were tired, and it was nearly dusk when he reached the old camp where he tethered the mare and rode Niggar out on the tidal flat to take up the *Prototherium* fossil he had come to claim. He broke it loose just in time to escape the onrush of the tide, only to discover he had only about one-half of the skeleton with forelimbs and feet. Tomorrow, he would have to return and, with luck, find the rest.

Back at his little camp he picketed Niggar, took three boards he and his party had left on their previous trip, spread two dry blankets on the boards, and piled on two damp saddle blankets. He ate another cold supper and stretched out on his improvised bed.

The next morning he was up at 6 a.m. A cold wind had come up and he had not slept well; but it was time to get on the tidal flat. He could not find where he had taken out the *Prototherium*, but located two similar skulls and lower jaws close to each other, as well as a pair of lower jaws of a third. He started to leave the flat when, close to shore, he found what appeared to be a complete skeleton of another *Prototherium*. The rock, coarse sandstone, contained concretions; and the tail and pelvis, with femur and limb minus some toe bones, were showing on top. The skull and parts of the forefeet were also nearby.

George felt sure this was a valuable find, and he returned with it to his makeshift camp. He moved camp to a safe shore spot in a large sinkhole where he leveled off some dirt, built a fire, dried out the bedding, and made a bunk. He cooked a meal, made a spoon out of a stick of wood so he could eat a can of cherries, toasted his cold biscuits on a stick, and then settled down to number his precious specimens and get ready for the incoming evening tide.

He spent another day and a half alone, working frantically on the tidal flat, and found armadillo and horse skulls and other less important material. He had numbered fourteen items from this little expedition and felt very smug and satisfied. He packed all his gear, loaded his specimen bundles, and headed back to camp.

In later years George loved to talk about the expedition to Patagonia, perhaps because life was so different there and he knew few people had the opportunity to have such experiences. In "Thrills in Fossil

Hunting," a 1930 article published at Fort Hays State University, he vividly described one lonely trip out on the ocean floor:

> One day when the tide was at its lowest ebb, I had followed it to its very limit to find the finest skull of a large *Nesodon* we had so far discovered. Water was already covering the specimen, and knowing how useless it would be to try to dig it out, I left it and returned the next day prepared to collect it. But much to my surprise only the very top was uncovered by the water. I had waited too long and now would have to wait a whole month when the receding of the highest tide would again uncover it.
>
> I hated to leave my prize, but there was nothing left for me to do but go. A month later I arranged for two saddle horses, one to carry my bed and food and to bring back my specimen, and the other as a mount. I arrived on the spot and knew that far out there my prize would soon be uncovered by the fast receding water. I tied one horse to a bush along the shore and mounted the other one, followed the tide as it receded far out over the shallow ocean floor. Finally I saw the top of my find. Soon it was entirely uncovered and the water receded several yards beyond it. My horse turned his tail to the chilling damp wind and stood facing me. 'What a splendid companion is a faithful horse when one is out alone,' thought I.
>
> I removed the skull and rolled it from the hollow from which it had been dug and getting it beside my horse I started to load it but could not raise it above my waist. Finally, in despair I remembered a small piece of a broken life boat which had been washed up on shore. I rode at all speed to shore, tied my rope to the broken boat and dragged it out to the specimen. There was nothing to guide me to it, and in the darkness I could not see it from my horse, We followed the water's edge until finally my horse stopped. Not two steps from us in the very edge of the water was my find. I soon had it strapped on that battered piece of boat and safely towed to shore. The next day, by getting my horse close to a small bank I loaded it on his back and took it back to camp. It was then packed and shipped to the Field Museum.[18]

By the middle of April, each morning brought a cold wind and frost on the ground. Strong gusts staggered not only the men, but the horses with their packs. It was almost time to move north, away from the bitter Antarctic winter fast approaching. Work continued, however, in the Santa Cruz province until the end of May. The party collected 282 specimens of fossil mammals, together with a few specimens of fossil birds from the Santa Cruzean formation. This number included 177 skulls and a few skeletons, more or less complete. According to field notes, the collection included thirty-two genera of fossil mammals and a considerably larger number of species.[19]

For reasons he did not record, George marked an X on a building on this Comodoro Rivadavia postcard he sent to Mabel on June 24, 1923.

The party moved northward to the vicinity of Comodoro Rivadavia, and for the next two months they collected Recent mammals and birds as the weather permitted. Then they set out to reach the Río Chico and the locality in which the Amherst Expedition of 1911 had found excellent fossils of *Pyrotherium*, a Lower Tertiary tusked ungulate almost as large as an elephant. Their route took them across the pampas, 2,000 feet (600 m) above sea level, where there had been a recent heavy blanket of snow. After spending two nights on the high pampas, they reached their destination and set up a single tent where thorn-bush barricades served as a shelter against wind and snow.

Fossil discoveries were limited since the Amherst group had left few specimens uncollected. When a new supply of plaster of Paris arrived, the fossil hunters were able to take out a rare specimen of a fossil bird and some other specimens. But after only a week they moved westward to the region of lakes Colué Huapi and Musters. Finding the valley of the nearby Río Chico impassable, they had to return to Comodoro Rivadavia and approach their destination from the south. Rains halted operations until October, when they were ready to venture out again.[20]

Abbott went prospecting one evening and came back excited about the discovery of bones of large dinosaurs in the gray shale, apparently the target of previous collectors. The party went to the location the next morning to find bones dug out of the hillside and piled up in heaps on a level area at the base. Leg and pelvic bones and vertebrae— mostly broken—had been organized with the fragments of each bone constituting one pile. There were no wrappings, no marks to indicate who had done the work and laid claim to the bones, which totaled 3,000 to 4,000 pounds (1,350 to 1,800 kg) in all. Rain had washed mud from the hillside over the bones, which, also weathered by wind, sun, and rain, were badly decayed.[21]

A few days later the men found a deserted camp near the old fossils where a ring of stones marked the outlines of a bush-shelter, or tolda, such as Indians of the region had built at one time and shepherds still used for temporary camps. They found a drift pick, shovel, and two hammers, all bearing the mark of a Sheffield toolmaker, as well as some other objects. The shovel blade was rusted through, and the ashen handle was old and weathered. Obviously, the tools had lain there a long time. Neither graves nor other marks indicated any tragic event, and none of the old settlers in the region could offer any clue or explanation, except to say that some fossil hunters had been there about twenty years earlier.

In the outcrops of the Deseado formation at Lago Colué Huapi, Riggs, Sternberg, and company found promise of fossils and established camp on the lakeshore in an adobe house near the home of an Italian-Argentine fisherman. They obtained horses and found an old cart track that they repaired sufficiently to make it passable. They brought in their truck and used the road for eight or nine miles (13 to 14.5 km), then continued on foot.

The Deseado formation was about six hundred feet (90 m) thick, consisting of interbedded clay and sandstone. In the upper part of the formation they found teeth, tusks, and various bones of the huge mammals *Parastrapotherium* and *Pyrotherium*. From the molar teeth of the *Parastrapotherium*, which had a grinding surface of more than three inches square (19 sq cm), the broken jaws and the thirty-inch (76-centimeter) tusks, they could visualize the size and character of this great animal. At the time, no complete skull of the *Pyrotherium* had ever been found—only fragments.[22]

Toward the end of January 1924, the expedition moved westward to explore the terrain surrounding the San Bernardino Mountains west of Lago Musters, where the fossil beds were in the Cretaceous rocks. The party found no mammals, but did find dinosaur femurs— well-preserved leg bones weighing nearly half a ton (450 kg). They collected a few, but reminded themselves they were hunting mammals, not dinosaurs, and went on about their primary assignment.

A little later they found a fossil forest. The first indication of its existence had come when someone brought in a fossil pine cone, saying it had been found 180 miles (288 km) to the south. Shortly thereafter, they received two similar specimens, said to have come from sixty miles (96 km) to the west. When someone brought other specimens reportedly from the south, Riggs hired him as a guide; and the group traveled four days by car to reach the site. It had been three months since they had seen the first pine cone. High up, in Sierra Madra y Higa, they found fossil trees, some with stumps standing, others prone with broken branches and cones scattered about. It was a forest of *Arecaria* or Brazilian pines, preserved where they had grown. Riggs, Sternberg, and Abbott took up 150 specimens of cones, twigs, and branches.[23]

In May 1924 the party broke camp and headed north, away from the bitter winter. For Sternberg and Abbott, it was the end of the expedition. They returned to Chicago with Riggs, who, after a short report interval, went back to South America, primarily Bolivia, in honor of his commitment, which continued until September 1927.

In summing up the two expeditions, Riggs reported that the fossils collected derived from no fewer than twenty-two different sites. Fossil invertebrates had come from strata of Cambrian, Miocene, and Pliocene age. The parties had examined most of the formations of Argentina and Bolivia at that time known to yield fossil mammals and had taken a series of more than 1,200 photographs to record the work of their expeditions, as well as to illustrate subjects of more general interest.

The Argentine and Bolivian governments allowed all of these collections—except a certain number of duplicated specimens—to be exported to the United States. The shipments totaled nearly three thousand specimens of fossil mammals, birds, reptiles, mollusks, and plants, all of which were sent to the Field Museum where it took several years to study and prepare them for exhibition.[24]

Each fossil hunter, of course, came away with some personal mementoes of the expedition. George had the pretty shells for little Ethel and he had also obtained an ocelot—or at least its salted-down pelt. A sheepherder near Puerto Deseado had found the animal in his flock, killed it, and given it to George, who skinned it and salted down the hide in order to preserve it. Later he sent it to Denver to be mounted, and it became one of his most cherished souvenirs. One of the very few items from the expedition he was able to keep, the ocelot found a final home in the Sternberg Memorial Museum in Hays.

CHAPTER SEVENTEEN

Back Home on the Range

As George Sternberg neared Chicago at the end of his long journey from Patagonia in May 1924, he faced an uncertain future. With his brothers and father now scattered and working independently, he had no home base. And his wife's letters to him in Patagonia, as erratic as they were disconcerting, had given him reason to fear for the unity of his own family. Soon, he decided, he would make his way to San Diego, the closest thing he had to a home now that his wife, children, and parents were living there.

In Chicago George quickly completed his business with the administrators at the Field Museum; and, after receiving their assurance that they would be interested in knowing where he eventually settled and in perhaps buying future fossil finds, he went to Canada to see his brothers and to tidy up some unfinished matters. Then he headed for California. He'd never been away from his parents for so long. In fact, before his trip to South America, he'd never been separated from his father for more than a few months at a time since he had been old enough to start working regularly with him. He wanted to discuss some long-range plans with his father, and he wanted to find out what had become of some of the fossils he and his father had stored for years in Lawrence.

George went by train, stopping briefly in Lawrence. All his relatives were gone, and the old workshop, or laboratory as his father proudly called it, was closed. He did, however, locate numerous boxes of stored fossils—some ready for display, others still in their plaster of Paris cocoons. He knew that eventually it would fall to him to dispose of them. He went on across Kansas, enjoying the lush spring beauty of vast fields of green wheat waving in the wind. Grain elevators, like lighthouses, stood as sentinels at every town along the railroad. He had been away from Kansas for about sixteen years. The unbroken prairie he remembered had been transformed into cultivated farmland, but herds of cattle nibbled at the buffalo grass in pasture lands enclosed with fences of barbed wire secured to post-rock fence posts. Busy little towns hugged the path of the railroad, and the old trail ruts of pioneers' wagon trains had given way to the beginning of a highway system parallel to the railroad across the state. George knew he had come home. Just where in Kansas he would settle, he wasn't sure, but it would be out west, near the chalk beds. This had been Mabel's home country, too. Maybe she could be swayed to come home with him.

The train emitted a long, moaning whistle, slowed for a little town, then sped westward—on through Colorado, across the Great Divide, across the desert and finally to California. With the passing miles, George remembered stories his father used to tell around the campfire about harrowing experiences and valuable fossils he had taken out of the rocks. George had hunted in several Western states himself and had no doubt that in the years ahead he would retrace his father's trails and blaze new ones for himself, finding more fossils.

Monument Rocks in Gove County (Kansas) dwarf men and machine.

He kept thinking about how little the people of Kansas—and in other states as well—knew about the fossils and evidence of prehistoric life to be found almost in their own backyards. He wanted to make school children aware of the presence of fossils, to awaken their interest in the rich natural history they took for granted. He was confident his father could help him formulate plans. George and Charles H. had always enjoyed a close relationship, so close, perhaps, that he had excluded Mabel from his inner thoughts and purposes in life. He hoped it was not too late to make amends to her.

At last in San Diego, George told Charles his plans. Charles listened attentively and offered encouragement. Mabel, however, was less than responsive, and, after a short visit, George was ready to go home to Kansas. Mabel refused to accompany him, but agreed to let young Charles W., a lad of sixteen, spend the winter with him. George's daughter Ethel, now thirteen, would remain with her mother.

George (center) with Charles H. and Anna Sternberg in San Diego.

George decided on Quinter as a home base. It was a bustling little town on the Union Pacific, north of the chalk beds in Gove County. He and his son moved into a private home where they could also arrange for meals; but since the weather was still warm within a few days they established a camp on Hackberry Creek close to the chalk beds. It was October, too late for young Charles to attend the fall term of school. School could wait until the spring semester. Life had run full cycle for George. He was doing just what his father had done with him a generation ago!

Sternberg lived and worked upstairs in this abandoned Oakley schoolhouse.

As winter approached, father and son moved into Quinter so young Charles could enroll in high school. This arrangement, however, was only temporary; and before Christmas they moved again, this time to Oakley about thirty minutes west. They set up housekeeping on the second floor of an abandoned schoolhouse. George designated part of the space as his laboratory and storeroom. The downstairs served as home for the school mechanic and as a garage for school vehicles.

George had never needed elaborate housing, but this was more primitive than he would have chosen had he been able to afford better. Young Charles had not been thrilled with the decision that he leave California and move to Kansas with his father. He sometimes felt pressured to join his father hunting fossils, but other times enjoyed it, even though he never shared the passion for the work chosen by his father and grandfather. He consoled himself that at least he was now in a good school and could expect to be there for at least the rest of the academic year. They had a roof over their heads, and George soon felt at home.

Sternberg wished he could qualify as a teacher so he could work with children on a day-by-day basis, cluing them in to the wonders hidden in the rocks and wastelands. But with only a fifth-grade education, he knew his teaching would have to be out in the field rather than in the classroom. With his father, he had discussed some ways he could stimulate the interest of the general public in fossils. He envisioned fossils hanging in banks, libraries, and schools—places people visited almost every day. He anticipated inviting teachers to bring their classes to meet him in the chalk beds and spend a day hunting sharks' teeth, the easiest fossils to find and surely the most plentiful. Perhaps a child might discover a bone fragment of an ancient skull. His ideas were unlimited, and his plan seemed workable.

George contended that the growing population in western Kansas had little knowledge about either the ancient Cretaceous sea that had covered the region millions of years ago or the animal life that abounded then. Some of the world's richest fossil beds lay just a few minutes' drive from Oakley. Only a very few people, he believed, had ever visited the area or looked for fossils.

As he formulated his plans, he wrote about them to his father and to his longtime friend Dr. C. W. Gilmore at the Smithsonian Institute. Both answered with encouraging letters. Gilmore wrote:

> Your project of providing the Kansas High Schools with collections of fossils representative of their respective localities strikes me as a highly commendable idea and one that should receive the heartiest endorsement from all those who are interested in furthering the science of vertebrate paleontology. The idea of making relief mounts of the specimens is especially good for school purposes as it not only makes the fossils intelligible to the layman, but it renders it possible to utilize otherwise unused wall space for their storage. They are thus always available for study and out of danger from careless hands.
>
> I wish you success in your venture.[1]

Gilmore also reminded Sternberg that, although the Smithsonian did not have "much money for the purchase of specimens," the museum did "manage to secure an exhibition piece every now and then." He said he hoped George would keep him in mind for any large museum specimens he might have.

From San Diego, Charles H. Sternberg added his blessing to Gilmore's, almost as if he were finally relinquishing parental control and granting permission for George to go ahead on his own. George kept his father's letters in his files for the rest of his life. One, an open letter of introduction, read:

Feb. 20th, 1925

TO WHOM IT MAY CONCERN:

> My son, George F. Sternberg, has written me, he has conceived the plan of making representative collections of the famous mosasaurs, fishes, plesiosaurs and pteranodons, from the richest fossil field in the world. The chalk of Kansas. This material [is] to

be beautifully prepared and mounted for the use and study of the High Schools, of Kansas. This field I worked for 20 years and sent the material over the civilized world wherever science is appreciated.

It is fitting that my son, who since a boy of fourteen years of age has assisted me in my life work, finding some of the finest skeletons of extinct reptiles and mammals that now are among the chief treasures in the large museums of the world, should make this notable attempt to interoduce [sic] the extinct animals of Kansas into her schools.

He has been connected with some of the best museums in Canada, and the United States. As an assistant in the Victoria Memorial Museum in Ottawa, Canada, he has collected and prepared some of their finest dinosaurs and other animals. He labored under the greatest of living paleontologists, Prof. Henry F. Osborn, President of the American Museum of Natural History, New York City, N.Y. He has quite recently spent months in the South American fossil fields for the Field Museum of Chicago, Ill.

I consider his purpose laudable and hope this Born Fossil Hunter meets with well merited success.

Faithfully,
s\ Charles H. Sternberg[2]

Early the next spring George and his son headed for the chalk beds southeast of Oakley and began what turned out to be a productive summer of hard work. George always carried a small notebook in his pocket, and his penciled notes provide the best accounts of his discoveries. Of a marine turtle, he wrote:

May 23, 1925 3 mi. SE of new store at Old Elkader. The specimen lies bottom side up. There seems to be no chance for a skull nor front feet. A great deal of the front part of shell seems to be missing. No tail. I hope to be able to make a good shell out of this, but the rock is very soft and some exposure is near the bones . . . The shell is about 45" [114 cm] long by 37" [94 cm] wide with lower shell pushed slightly forward. . . size of frame 51" [130 cm] x 43"[109 cm] x 5" [12.5 cm] — 1 pkg of fragments. After taking up this specimen I have decided to open the side that was down first. This will be the top of the shell and after it is cleared I can put on a false base and turn it over.

Across the penciled page he later wrote: "Sold to U. S. National Museum, March 1926.[3]

About two weeks later, while working in the same area, son Charles W. and his father found one of their prized discoveries—a nearly complete skeleton of a twelve-foot (3.6-meter) *Portheus* with a smaller fish in its stomach. The skeleton of the smaller fish, which George

Twenty-five miles from the nearest railroad and town, Elkader Post Office also served as a gasoline station, grocery store, and gathering place for scattered homesteaders and fossil hunters.

identified as a *Gillicus*, was broken as though desiccated or partly digested. When it was whole, George figured, it probably measured nearly four feet (1.2 m) long. He described his prize thus:

<u>Specimen #6-25</u>

The ends of the jaws were sticking out of the rock and broken down. All teeth are weathered away and the top of skull or side up is badly damaged, but the other side should be fine or very good. I can restore the teeth. The large fin just back of the skull is good. Vertebrae except 3 or 4 just back of head are in place all slightly turning up towards the back so that the fish lays as is so often the case in a twist. All fins are present. Missing parts can be restored. The tail lobes are complete and measure 41" [101 cm] apart from tip to tip.

I am taking this up in 2 or 3 sections and will straighten out the tail to go into a frame 45" [114 cm] wide. I am putting the plaster tight onto the bones so they won't slip, therefore we can only present the left side, which is down. This should make a very fine slab-mount specimen.

Remarks: I told some 15 or 20 people about this specimen Sunday and Saturday in Oakley and there was over 30 cars out on Sunday. Wagons and riders to see the specimen. I believe over 150 people. There were 17 cars here at one time and they kept coming till dark. Monday people came all day long. At least 37 people were here. Two loads came from Scott [City] after sundown and we left the quarry by the lights of the cars.[4]

217

A large fish is encased in frames and sealed with plaster while curious neighbors watch in the Logan County chalk beds.

Across the penciled page he wrote "Sold to the U. S. National Museum, Mar. 1926."

This discovery, of course, attracted the attention of the press across the state, and major newspapers carried the story.

Just a few days later George found another *Portheus* skull, which he identified as #11-25, in the same general area. This one, he believed, was the right size to help complete his spectacular Specimen #6-25. "The left side was down as found," he noted, "so would work OK with No. 6, if it is the right size. ... Have taken it up in one plastered section. No frame." He sent it to the National Museum marked "free-gratis."

By the end of the summer of 1925 father and son had found and removed thirty-two specimens. They were sold to the State Museum of Illinois at Springfield, the Kansas State Teachers Colleges at Hays and Pittsburg, University of Colorado, Washington University at St. Louis, and the Oakley (Kansas) Consolidated Schools. Within a year other specimens went to the Museum of Comparative Zoology, Cambridge, Massachusetts; a large slab of crinoids to Dr. O. T. J. Mortensen of the Zoological Museum, University of Copenhagen, Denmark; and many mammal specimens to the California Institute of Technology.[5]

In 1926 officials of the Union Pacific Railroad, which carried the Sternberg specimens out of Oakley, invited George to write an article about his fossil hunting for their magazine. In simple language he presented the story of the great fossil fields of marine life not far from the railroad and Oakley. Of the local landscape he wrote:

> You can enjoy this type of scenery. First you visit the pyramids which lie 26 miles [42 km] southeast of Oakley. They stand out on the prairie towering 60 feet [18 m] or more above the ground. They are all of sedimentary deposit and are carved out by wind, frost, and rain. They remind one of the ruins of an old castle. The most northerly one is probably the most striking example of a human face ever carved from the rocks by nature.[6]

In the article, George explained that although he had worked for about twenty-five years, going as far north as Canada and as far south as Patagonia, he always returned to the Kansas rocks to wonder why he ever left them for other fields.

One of his finest specimens, he wrote, was "the 13½-foot [4 m] fossil fish, *Portheus*, which has been purchased by the Oakley citizens and schools for exhibit in the main corridor of the grade school." He added: "This is only the fourteenth specimen of this great fish ever to be collected and the second one to remain in Kansas. Furthermore, it is

This fine thirteen and a half foot (4 m) Portheus *was found in Logan County and purchased by Oakley citizens for their school.*

219

George Sternberg lectures to Oakley School children about the Portheus *purchased for their school.*

the only one to be permanently located so near its original burial ground. The specimen was in a splendid state of preservation and attracted national attention."[7]

Elsewhere in the article, George called attention to the fact that the Oakley school fossil was the same species as the *Portheus* with the small *Gillicus* in its stomach that he had sent to the U. S. National Museum the previous year. He also described the portion of a great flying reptile *Pteranodon* on display at the school. *Pteranodons*, which often had wing spreads of twenty-five feet (7.5 m) from tip to tip, soared over the Cretaceous seas in search of the fishes on which they fed. "They must have been strange-looking creatures," George explained, "with their bat-like wings, long slender heads, short legs and no tails." He described another fossil exhibited in the Oakley schools as a well-preserved skull of the mosasaur *Platycarpus*, with a lower

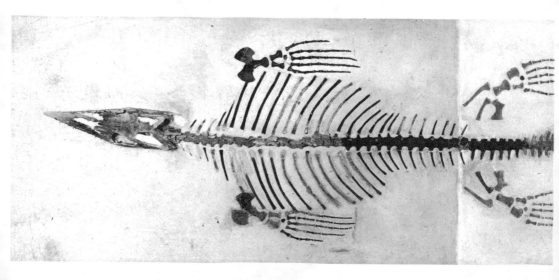

jaw twenty-one inches (53 cm) long and a body about sixteen feet (4.8 m) long.[8]

George was anxious to have his fossils displayed throughout western Kansas and exhibited them as an artist would his paintings or sculptures. A skull of the small mosasaur *Tylosaurus* was on display in a bank in Grainfield. The lower jaws of a young mastodon, with inferior tusks in place, attracted attention when shown in the high school at Healy, a short distance south of the pyramids and the fossil beds.

Gone was the great need for secrecy among fossil hunters, and George found pleasure in the opportunity to take classes on field trips. The youngsters often found sharks' teeth, stray shells, and other relics of the ancient sea. He spoke to civic groups and clubs, and maintained extensive correspondence. At the end of his working season he made up a list of all his available fossils, with a price for each, and mailed copies to all the museums with which he and his father had dealt. This campaign paid dividends; he was now building a reputation for his own discoveries and preparations—a gratifying turn of events.

The little community of Oakley was cognizant of the attention it received because of fossils. The superintendent of schools and members of the school board wrote a letter of appreciation to Sternberg in the summer of 1926, acknowledging that his work in the community had aroused interest as nothing else had done for years. The big *Portheus* had attracted fifteen hundred people to the Sternberg laboratory, and the school boasted six fine wall specimens and a case of small material of rare value. The letter called Sternberg's installation of this material in a school museum "an invaluable method of bringing a knowledge of … fossil life to the school and public at large" and commended him as a "man of character, scientific knowledge and ability in his line of work."[9]

This small Tylosaurus, *a sea lizard twelve and a half feet long (3.75 m), was found near Monument Rocks. The darker portions are reconstructions.*

As Sternberg's efforts to attract students became known, teachers and classes came from towns more than a hundred miles (160 km) away, and a geology professor from the University of Missouri in Columbia brought his class of eighteen students in a caravan of four trucks nearly five hundred miles (800 m) to go with Sternberg to the Logan County crinoid quarry. Closer to home, Dr. L. D. Wooster, professor of geology at Kansas State Teachers College at Hays, took his class to the same area and spent considerable time with Sternberg. Wooster was much impressed by Sternberg, as was one of the students, a graduating senior named Myrl V. Walker. Unknown to Sternberg at the time, this visit inspired the professor and laid the groundwork for a change for the fossil hunter, one that would net him a permanent base and position in a college environment.

University classes, as well as local residents, followed Sternberg to the Niobrara chalk fossil beds.

CHAPTER EIGHTEEN

Museum Peace

P rofessor Wooster wasted no time when he returned from his field
trip with George Sternberg. He wanted the man to move to Hays
and affiliate with Kansas State Teachers College there. It would be
prestigious for the college to have a resident paleontologist. Part of a
new building was available for a museum, and who better for the job
as the first curator than a man with an international reputation, one
who knew, probably better than anyone else, the treasures of the
western Kansas chalk beds?

The state of Kansas had made no appropriation for such a museum,
but the college offered Sternberg a contract to work on developing one
during the school year, with summers free to work for himself hunting
fossils and to sell any specimens found. Sternberg's salary for the first
year would be twenty-five dollars a month for nine months.[1]

Almost as excited as Wooster and Sternberg about the arrangement was the young graduate student Myrl Walker, who saw a unique
opportunity for himself if he could become associated with Sternberg.
Walker began teaching a few classes at the college that summer,
substituting for Wooster, who was seriously ill. The college purchased
a large *Portheus* from Sternberg to help him finance his move to Hays;
and Walker was on hand to help crate and move fossils and equipment
from Oakley.

That summer of 1927 Sternberg, Walker, and a student assistant
loaded camp equipment and working tools into Sternberg's panel
truck and headed for Wyoming where Sternberg had a contract to
work. North of Lusk they found a titanothere skeleton that Sternberg
later sold to the California Institute of Technology; a fossil horse that
was found by Walker and eventually went to the American Museum;
and several other lesser fossils that remained in Hays.

George Sternberg.
—Kansas State Historical Society

Writing reports of his daily activities was not one of Sternberg's strengths. He kept field notes, written in pencil in a small book, and generally this was his only written record of the work of a summer, except for what he filed with his sponsoring museum. Typical of his accounts of the problems of identifying and recovering fossils is this note describing "Specimen #40, found in the Craddock Quarry, Permian of Texas":

> Likely *Dimetrodon*? What seems to be the scattered skull with at least one complete lower jaw and no doubt the other. There is at least 20 vert. with more or less complete long spines not in place but rather scattered about in somewhat of a line, there are numerous limb bones and a number of them with spines have been placed on top of the skeleton before wrapping it with plaster and burlap. No care has been used in putting parts together and it will be necessary to wash the pieces and then associate them. There is little use to try and determine how much of this skeleton is present due to my lack of knowledge of them but if all the bones

wrapped in packages belong with this skeleton as I believe they do then I would guess that over half the skeleton is here. I fear however the feet are gone. I am taking most of the material up in one large plastered section just as the bones lay. There are 3 wrapped pkg. from the back part of what is here. There is very little rock on the bones. It will remove easily. This should prove a good specimen once it is cleaned up.

Later, he added this notation across the notes: "Sold U. of Calif. Jan 6 1930 to be Paid for in Aug. 1930."[2]

Altogether Sternberg found, described, and numbered fifty-one specimens in 1927. In the Florissant beds of Colorado, he found and recorded quantities of smaller specimens, both plant and animal. He recorded the choice fossil leaves on 120 green tags and counted them off in units of five. He used red tags to record fifty-five choice fossil insect specimens and tagged another group of poor insect specimens to bring his total to 255 specimens. He made no mention of how he disposed of this collection.[3]

Back at the college in the fall of 1927, Sternberg began his new job in earnest. Area teachers in increasing numbers brought classes to his fledgling museum in Hays, and when possible went with Sternberg to the Niobrara chalk. He arranged field trips for Wooster's classes to go to Ellsworth County, where the Sternberg family began hunting fossils so long ago, and to Castle Rock and the badlands of southern Gove and Logan counties. He rarely went hunting alone. The names of would-be student helpers kept full a list, from which he would choose a few for each trip. Sometimes he talked freely about where they were going and what they might be looking for, but sometimes he slipped back into patterns from that early period in his life when his father had required utter secrecy, as if others were waiting to pounce on his discoveries.

One of George's early student volunteers, Rudolph Barta, recalled sixty years later—in 1986—how secretive Sternberg had been on a trip they made to Ellsworth County. George wanted no information to leak as to where they went or what they found. Actually, what they found were fossil sassafras leaf impressions like those his father had found as a teenager. Barta remembered hunting for peculiar oval or rounded concretions that appeared to be seamed and fit together like halves of an eggshell. Breaking them open revealed the perfect imprints of leaves.[4]

At this time, all the Sternberg fossil hunters were still active. George was eager for his family to see him in his college environment; and within a year after the move to Hays, his father and mother visited from California. Both of his brothers—Levi, then with the Royal Ontario Museum, and Charlie, with the National Museum of

Charles visits George in Hays. —Kansas State Historical Society

Canada—also visited and brought a few specimens from their own collections for the new museum at the college in Hays. All offered suggestions, of course, as to how the displays should be arranged and featured. George listened respectfully, and when they all went home, did it his own way! He was a craftsman of no small skill. For years he had made frames, chests, and storage and shipping boxes; and now, with the college supplying oak that matched the trim of the museum rooms, he was making beautiful display cabinets and cases.

When the summer of 1928 rolled around, Sternberg, Walker, and an assistant, Edwin Cooke, again loaded the panel truck. They took a tent for storage, but planned to sleep in the truck. For Sternberg, this was a normal way of life, perhaps made even a bit luxurious by an ice chest from which he could keep camp supplied with a variety of good food. Gone were the days of tinned sardines, salmon, beans, and hardtack.

Even late in his life, sleeping in the truck was often part of the routine for George F. Sternberg.

For Walker, the experience was a new one, but one he, as well as Cooke, came to love. Walker was destined to spend most of the rest of his life either in the outdoors or working around museums. His association with Sternberg would continue for the rest of their lives. That summer they headed for Wyoming and Montana, with George again working under Charles Gilmore, field supervisor for the Smithsonian in the upper Cretaceous near the Blackfeet Indian Reservation.

The summer pattern was repeated for several years. Sternberg and Walker were usually joined by Gilmore. The American Museum in New York contracted for vertebrate fossils primarily from the Eocene and Pliocene epochs. Sternberg sought to fulfill these contracts as well as to meet Gilmore's needs, but also kept his eye open for other possibilities, for other buyers for his own collection. The hunters found

Edwin Cooke (left), a student assistant, helps C. W. Gilmore prepare specimens on a summer expedition near Cut Bank, Montana.

fossil insects and leaves, as well as a small camel, mammoth, and dinosaur bones. Walker found, collected, prepared, and described a skull of the small lizard-like *Rhineaura sternbergi.*[5]

During the winter months at the college, Walker taught a couple of classes and continued his graduate study. He also assisted Sternberg as much as possible, learning how to prepare and mount specimens; and together they learned how to run a museum.

After a year off to complete work on a graduate degree at the University of Kansas, Walker returned to Hays, teaching full time. Sternberg scheduled more field trips: one of his most-publicized weekends was a trip in April 1930 for a group of Kansas City

George F. Sternberg with some trophies from the season of 1928.

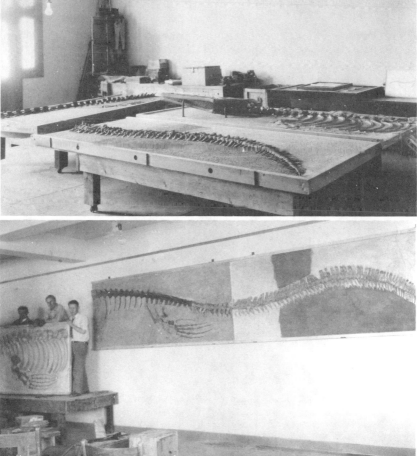

1. *Preparation of a large fossil takes time, detailed work and space. George Sternberg works with a paint brush to remove fine particles from this three-slab specimen. 2. The head and two sections of vertebrae are finally ready for display. 3. Helpers lift the head section of the mosasaur to complete the specimen at Fort Hays State Museum.*

professional people, mostly doctors and lawyers, with a sprinkling of sports announcers and officials. The story in the *Kansas City Star* carried the headline, "George Fishes with a Pickaxe!"[6] A group of Hays physicians had organized the party and joined the group as it headed for the chalk beds of Gove and Logan counties. They found no important fossils, but that was relatively immaterial. It was good publicity for Sternberg and his budding museum.

George's son, Charles W., graduated from Hays High School in 1928 and enrolled at the University of Kansas, taking courses in geology. He had no passion for the fossil fields and now was able to remove himself from any obligation to accompany his father on further expeditions. However, he did not divorce himself completely from his father's work, and occasionally went on trips with him. Later, Charles earned a master's degree in geology at the University of Chicago and became a geologist for the United Nations in Israel and Turkey.

Life was good for George Sternberg at this point. His arrangement with the college was flexible, and he foresaw the probability that Hays

Sternberg urged businessmen, teachers, and students to try their hands at digging for sharks' teeth or other fossils. —Sternberg Museum

would be his permanent base. He was lonely, but was in a college community that afforded many opportunities for social involvement. Lack of formal education prevented him from achieving faculty status, but he was a favorite among faculty men who included him in their club and weekend outings: several of them became somewhat regular weekend assistants for Sternberg and Walker. All he really needed for complete fulfillment, he felt, was a wife. Mabel had obtained a divorce in 1926.

Ever loyal to his religious upbringing and heritage, George was active in the Lutheran Church of Hays, where he met a woman as lonely as he. She was Anna Reed Zeigler, a widow with teen-age children, about the same ages as George's son and daughter. They married in October 1930.

For the next thirty-five years, George had the companionship and lifestyle that he had never been able to enjoy with Mabel. His long periods of wandering were over. Now his wife was willing to accompany him occasionally on summer trips, and for the rest of the year they could enjoy a normal home life and community activities.

Sternberg had been selling items to the Smithsonian, but in 1933 that market shut down: no money for fossils! Sternberg and Walker

Anna and George Sternberg in 1952.

sought other markets. While teaching in Clark County, Walker wrote to Childs Frick of the American Museum in New York, asking if it could finance an expedition. Frick wanted someone to work in northwest Kansas on a limited, part-time basis. But by the time Walker received an answer, he had aligned himself with the National Park Service. So he wrote again to Frick, recommending Sternberg for the job. This led to a continuing summer job and happy relationship for Sternberg.

The museum had grown faster than Sternberg or college President W. A. Lewis had dared to hope. The Kansas press was kind to Sternberg and his many expeditions, especially when they involved taking large groups of high school students to the fossil beds. The people of western Kansas always felt a personal interest and pride in the college at Hays as their own school. Now they were becoming aware not only of Sternberg and his fossils, but of the fact that the fossils were being prepared and exhibited close to home.

The school, as well as the museum, grew rapidly. When the liberal arts program was accredited, the school was rechristened Fort Hays Kansas State College. The new name capitalized on the school's historic heritage: its first buildings were part of old Fort Hays, and the first campus was the fort's parade ground. Of the seventy-six hundred acres (3,000 ha) of land held by the fort, more than four thousand acres (1,600 ha) had been designated in the gift to the state of Kansas for establishing the college.

Myrl V. Walker looks up at the mosasaur on display above the cases of smaller fossils at the college museum.

Sternberg became keenly aware of his academic weakness, but his forty-plus years of working in the field and with paleontologists had added immeasurably to his education. He occasionally had opportunities to write of his discoveries and adventures, as his father had done before him. Fortunately, by now he had a part-time student secretary, and his good friends, Walker and Wooster, quietly edited his copy so that the printed word never embarrassed its author.

The economic tragedies of the Great Depression of the early 1930s were compounded in western Kansas by the unbelievable dust storms that ravaged the Great Plains from late in the winter of 1934 to the summer of 1937. Massive billowing clouds of dust, black as the darkest night, rolled across the horizon, bringing choking darkness in midday and paralyzing activity. Schools opened in the mornings as scheduled, but often closed quickly when word was sent out that the dust was rolling in. Most growing things in the fields were uprooted. Rabbits moved into town to eat leaves and bark from trees. Great drifts of pulverized earth buried farm machinery and fences, and caused livestock to choke to death or die of dust inhalation. Dust pneumonia was not uncommon among farmers who fought desperately to keep their stock alive.

Fossil hunting in Kansas was, of course, almost at a standstill. Barnum Brown, who had grown up in Kansas half a century earlier, sat back in his chair at the American Museum of Natural History and read of the red dirt rolling north from Oklahoma and Texas into Kansas, mixing with the black dirt blowing in from Colorado and Wyoming. He perceptively commented that the blowing dust was a boon to fossil hunters because it would uncover countless fossils that otherwise never might be found. While this was probably true, the average victim of the Dust Bowl would have drawn little solace from such a statement.

It took years for the nation's economy to recover from the "Dirty Thirties," and even longer for the Great Plains to recuperate. As Brown had suggested, the fierce winds had swept away tons of chalk rock and earth, and laid bare whole new faces to the cliffs, pinnacles, and plains. When the winds finally subsided and the rains came and the dust settled, there indeed were new fossils in sight. Even a small portion of a bone was enough for Sternberg's eye, trained to spot the flecks of black coloring that often indicated the remains of such things as jaws and fins. He might chip just under the surface with a pickax and find scales, fins, or vertebrae, perhaps as small as a dime, but leading to more.

Every rain and every wind storm exposed more fossils when the powder-fine chalk rock eroded. Once exposed to the elements, most fossils tend to weather away and disintegrate, sometimes in less than

Edwin Cooke struggles for secure footing while digging in the Oligo-cene of Nebraska.

a year. After the dust storms subsided, Sternberg felt he had a whole new world to explore. Sometimes it seemed fossils were just waiting to be discovered, and now many local farmers and laborers were motivated to look for old bones, since there was an authority nearby and area museums where the discoveries could be prepared and exhibited.

Sternberg (right) and helper lie down on the job—for detail work with small tools and soft brushes—in the Swayze quarry in Clark County.

Clark County, Kansas, contained a small gold mine of fossils, and a local man, Jack Swayze, became the unofficial collector for the county. He worked five summers with Sternberg and sometimes with Walker.[7]

A road construction crew working in southeast Trego County in 1932 came upon some unusual bones and summoned Sternberg, who removed and restored them—the lower jaws of the prehistoric elephant, described by Walker as *Amebelodon*, "a mud-grubbing mastodon that lived and fed at the edge of shallow swamps." The jaw, more than five feet (1.5 m) long and shaped like a giant shoehorn, caused the species to be called "shovel-tusked." The molars measured eight and a quarter by four and a quarter inches (21 by 11 centimeters), and the upper tusks were almost seven feet (2.1 m) long and eighteen and a half inches (47 cm) in circumference at the base.[8]

Sternberg's work in classifying and preserving this specimen earned him recognition in a monograph by Dr. Henry Fairfield Osborn and won the specimen its permanent exhibit in the museum at Fort Hays State College.

The dust did not blow every day, and Sternberg, like all other workers, followed his usual routine as much as possible. One day late in 1934, he and his assistant and former student George Pearce were on the ranch of W. C. Stout on the Cimarron River, northwest of Liberal in the extreme southwest corner of Kansas, where they found

Sternberg prepares upright elephant bones for removal.

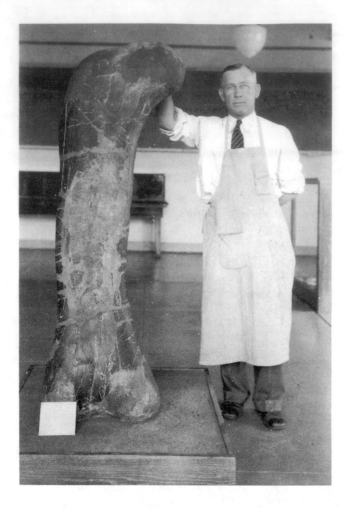

Sternberg poses with a dinosaur femur.

what Sternberg identified as a "fill-in" of an ancient lake bed. "Fill-ins," according to his definition, are usually of fine sand, marsh grass, and vegetation holding an abundance of shells and animal bones. This bed yielded Sternberg twelve hundred pounds (540 kg) of bones, including parts of seven different mammoths, three mammoth skulls, the partial skeletons of a fossil horse and a fossil bear, and parts of small wading birds.

Whenever the opportunity presented itself, Sternberg liked to feel he could do a bit of teaching as he visited with students. When reporters for the college newspaper asked him about the discoveries at the filled-in lake bed, he explained that the different elephants evidently had waded into shaded marshy places and died. Then, their bones had been trampled into the mud by other elephants. Although no complete skeletons were recovered at the lake bed, the most exciting specimens were the hind legs and one front leg of a mastodon that he had found upright and in place, just as they were when the animal was alive. Sternberg could only conclude that the elephant had died in an upright position after having mired down in the marsh. The two hind legs were standing five feet (1.5 m) apart.[9]

On the way back to Hays, the hunters stopped in Gove County where they found a rare peccary, an ancient pig-like animal that grew to about three feet (91 cm) in length. This trip was the last of three Sternberg made for Childs Frick in 1934, during which he logged approximately fifteen thousand miles (24,000 km) and covered areas in Idaho, Colorado, Wyoming, Utah, and Kansas. George's son Charles W., now out of the University of Kansas and not yet permanently settled, joined them for part of the trip, then went with his father to Canada that fall to complete some research and work George had left unfinished when he went to Patagonia.

In the late 1930s George continued his pattern of spending most of the winter months in Hays, working the summer season for Gilmore or Frick, traveling to the Judith River in Montana and Big Horn Basin in Wyoming, then going to New Mexico and Arizona with Dr. C. Lewis Gazin of the Smithsonian, searching for mammals in the San Juan basin. When working apart, Sternberg and Walker corresponded regularly; and whenever and wherever possible they met for even a few days or weeks in the summer.

George's financial picture brightened: his salary at the college increased steadily; his sales improved as the economy recovered. And with the extra cash, he was able to invest in rental property near the college.

In 1938 George and his helpers found tusked mastodons and other valuable fossils on the George Abell ranch in Clark County. These went in a shipment of seventy-five hundred pounds (3,375 kg) to the American Museum. Walker often spent his vacations from the National Park Service back home in Clark County, and Sternberg timed his trips to that area to coincide. Together they helped private collector Jack Swayze develop his holdings into a county museum in Ashland near the Kansas-Oklahoma Panhandle border. George Sternberg was at his happiest when he was helping new-found friends appreciate the discoveries at their own back door.

Sometimes his search for ancient bones was interrupted by observation of animal life of the present. One day, while walking toward a rocky area in Clark County, he stopped suddenly to watch a prairie dog. The little fellow stood on his haunches, with ears up and body erect. His eyes were fixed on something moving directly in front of him. Then George heard the unmistakable sound that identified the intruder as a rattlesnake. The snake lifted its head, then struck out at the little prairie dog, only to see the rodent jump back and disappear in his mound. Seconds later, he reappeared, and the scene was reenacted. The day was hot, and the snake could stand only so much heat. Gradually his movements slowed down, and finally the little prairie dog seized his golden opportunity and, with a quick lunge, bit into the base of the rattler's head, breaking its neck. George made sure the snake was dead and that there were no others around. Then he went about his fossil hunting business.[10]

Sometimes the fossil discoveries were not bones, but tracks of ancient animals, now fossilized and permanent. At an old watering hole in Graham County, northwest of Hays, Sternberg and a party of summer assistants found a large bed of tracks, which they identified as those of an elephant, camel, and rhinoceros. The elephant tracks were eighteen inches (46 cm) in diameter. Sternberg said this was one of the largest such groups of mammal tracks ever discovered in one small area in that part of United States.

Wilda and Myrl Walker and Anna Sternberg (right) marvel at the large fish they recently uncovered.

During the summers of the early 1940s, Sternberg and a small party worked west of the Continental Divide, sometimes going as far north as southern Montana, or as far south as New Mexico and Arizona. Sometimes he worked for Frick and the American Museum, sometimes with Gazin and the Smithsonian. But he always had a sponsor to finance the expedition. When on occasion Walker and his wife joined the party, Anna Sternberg also made the summer trip. Living conditions in camp were not as primitive as a decade earlier or in George's youth, but he was never extravagant in putting together his equipment. He never spent beyond the bare necessities for summers in the field.

The *Kansas City Star* once quoted Sternberg as saying: "It's a very hard life and sometimes pretty discouraging. For days at a time we may not make a find of any value. We always have to live in isolated regions without conveniences that many think are necessities and it's downright hard work. But I love it and wouldn't be satisfied doing anything else."[11]

The work was not only unpredictable, but sometimes dangerous. One day, as George paused to take a drink of water from his canteen, the side of the cliff where he was standing gave way and toppled into the ravine below, taking with it his tools and burying them under tons of dirt and shale. The spot where he had been working only a moment earlier was gone forever. He had to replace his tools, of course, but was thankful he had been able to jump away from the edge and escape uninjured. All else seemed insignificant.

In the summer of 1943, near Douglas, Wyoming, George was alone, working along the edge of a cliff when a rampaging steer headed in his direction. George scrambled up higher and grabbed for a clump of sagebrush growing on the face of the ledge. His feet dangled and his hands ached from the struggle to hang on to the wisps of vegetation. Fearing the sage would give way at any moment, he prayed he was higher than the steer could climb—or his horns would reach!

Off in the distance a young rancher named Bill Eastman had been rounding up a herd of steers and had thought he had them under control when the one now confronting George had broken away and raced off along a curve and down the canyon. The rancher gave pursuit. The cliff became steeper; and the bottom, narrower. He saw the man clinging to the wisp of sagebrush and cut the steer away by placing himself between the animal and the trapped man. Sternberg let go of the roots and slid down the cliff to Eastman and his horse. He brushed himself off, stood up, and reached out his hand.

"I'm George Sternberg, and I'm collecting fossils for the American Museum." Eastman then identified himself, and they visited briefly. Then Eastman invited Sternberg to come to the ranch for a fried

Summer vacation meant more fossil hunting for the Myrl Walkers (center) and George Sternbergs. Douglas, Wyoming, was one of their favorite hunting grounds.

chicken dinner the next Sunday. Sternberg not only went, but also returned the following Sunday and several Sundays thereafter, as the summer wore on.[12] The men discovered they had many similar interests. Eastman was an artist with college training in museum management. That summer he was on his ranch making wax models of big game. He had just graduated from the University of Iowa, preparing for a career as a museum curator, and expected to use the wax models as part of the background material he would need later.

Sternberg went back to Douglas the following summer and visited again with Eastman, trying to persuade him to move to Hays. But World War II was raging, and ranchers were frozen to their work of producing food for the nation. Eastman was interested in Sternberg's proposal, but could not make the move until the war was over.

"George did a recruiting job on me and was quite successful at it," Eastman said. "I finally went to Hays to look at the college and what it had to offer in my area in graduate school. Then, when the war was over, I bought a small house from George, just off the campus, and went home to tell my wife we were moving to Hays, and I was going back to school to get my master's degree."[13]

Eastman, his wife, and young daughter arrived in Hays about 3 o'clock one afternoon in the summer of 1945 and pulled his truck and trailer of household goods along the street by the Sternberg home. George rushed out to greet him.

240

"Am I glad to see you right now! Just put all this in my backyard for now and help me awhile this evening. Tomorrow I'll help you unload everything over at your own house!"

Eastman was shocked. He hadn't expected to go to work immediately. But he did. He parked his truck, and the two men piled into Sternberg's pickup and drove west about sixty miles (96 km) to Quinter where George had a heavy crate holding the skeleton of the giant turtle *Protostega gigas*. It was ready to be moved, but too much for one person to handle. Eastman never did find out just why it had to be done that particular evening, but together they improvised a block and tackle and lifted the crate into the truck. The turtle measured six feet (1.8 m) from tip to tip and was one of the most complete specimens of this species found to date. George was understandably anxious to ensure its safe removal to Hays.

This 1940s' photo of George (right) and his helper illustrates how cumbersome a plaster-wrapped fossil can be.

As they worked, they saw a rising dark bank of what appeared to be clouds, but what they soon recognized to be much more dangerous. It was a prairie fire, heading straight for them. Racing to complete their job and get away, they worked furiously, then jumped into the truck and sped eastward. They had not gone far when they realized they probably would have to drive directly through the fire, but they managed to outrun it, racing only a half mile (800 m) along the edge of the advancing flames.[15]

Eastman said he never looked at the turtle in the museum in the months following his arrival without smelling smoke and remembering their race with fire. His contract called for him to devote only seven hours a week to museum work. The rest of his work week was to be spent doing academic work in the Biological Science Department. He began doing oil paintings depicting the animals of the museum in lifelike situations, and he also did miniature reproductions of many of the fossils.

Sternberg sagely predicted in a report to the *News Bulletin of the Society of Vertebrate Paleontology* that Eastman would "add some color to the museum," adding, "He is painting a group of fossil rhino *Teleoceras fossiger* for me now. I believe he will be a lot of help to me. His interests are museums, and I need someone like that."[16]

Through his connection with Sternberg, Eastman worked out a program with Fort Hays State whereby he could teach a little, help Sternberg, and work toward his master's degree.[17] Eastman and Sternberg were good friends, sharing laboratory space, but working independently most of the time. Each respected the habits of the other. Sternberg was "particular, professional, and precise," in Eastman's assessment. He was an accomplished cabinet maker. If he

Fossil turtle ready for display.

Bill Eastman, museum artist, poses with some of the models he created and painted for Sternberg Museum.

didn't have the right tools, he remodeled what he had or made his own. He wanted everything clean, neat, and in good repair. He had good tents, and when they showed wear, he repaired them professionally. For camp use he made a set of kitchen chests, including ice chests, and he selected food carefully. He was a good cook. He expected and received the same kind of meticulous performance from those who worked with him. Sternberg knew nothing about Eastman's artwork and was content to make no attempt to learn it. Eastman, on the other hand, found himself accompanying Sternberg and helping him physically whenever possible.[18]

Their personal schedules were quite different. Sternberg had trouble sleeping. He went to bed early and was up by four. He often read in the predawn hours, then fixed his own breakfast and was ready to go to the field by daylight. But despite his short nights, he could hold his own with any assistant in the field all day long. Eastman was more conventional and followed the usual work-day pattern.

George relaxes outside his tent.

One day in 1946 Sternberg approached Eastman in the laboratory saying, "I think I have found a pretty good skeleton fish, but I need some help." He had a habit of organizing work parties whenever he had something big and exciting that he couldn't handle alone. This time he rounded up a group of faculty friends and made arrangements for a weekend trip to the Delmar Roberts farm, fourteen miles (22 km) south of Buffalo Park in Gove County. He had found only about six inches (15 cm) of the fish's snout protruding from a rock, but nevertheless felt sure there was a great deal more to be uncovered. The fish, he had no doubt, lay stretched out horizontally.

"We took off about ten feet [3 m] of the overburden," Eastman recalled, "then we had to put aside the bulldozer and heavy tools and work more carefully. By Sunday evening when we had to go back to town, we were within four or five inches (10 or 13 cm) of the fish and knew we had a big *Portheus molossus.*"

Sternberg returned to the site alone for the next stage of work. He spent several weeks chipping, brushing, scraping, and chipping some more, until the whole fish—fifteen feet, two inches (4.6 m) long—lay exposed. As he cleaned the central area of the *Portheus* he discovered parts of a small fish, a *Gillicus*, spread upside down over an area of seven feet (2.1 m) midway in the body of the *Portheus*. The parts were scattered and incomplete, but he found enough to make species identification and to realize he had a valuable discovery. He had found other examples of a fish within a fish, as had others; but this was the largest and most clearly defined fossil of this type unearthed to date. The skull of the *Portheus* was in excellent condition with fine big teeth showing; and Sternberg could count eighty-three distinct vertebrae.

When Sternberg had uncovered the fossil bones to half their thickness, he called Eastman and his other volunteers back for another weekend of work. Together they constructed wooden frames in three sections, each four inches (10 cm) thick, sixty-three (160 cm) inches long, and fifty-six inches (142 cm) wide. Sternberg painted all exposed bone areas with shellac, coat after coat of it, to harden and preserve the fragile bones before they could begin to crumble and disintegrate.

After fitting the frames around this fossil, George filled them with several inches of plaster. When it had hardened sufficiently, he nailed a cover on top of each frame. Each section weighed approximately five hundred pounds (225 kg).

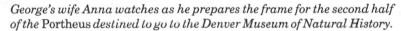

George's wife Anna watches as he prepares the frame for the second half of the Portheus *destined to go to the Denver Museum of Natural History.*

The next step was probably the most physical. The workers dug a trench six to ten inches (15 to 25 cm) deep under and around the frames, undermining them until they could lift them and turn them, first on end, then bottom side up. The final step was to load the sections into the truck and take them to the museum in Hays.

Sternberg spent most of the following winter chipping, scraping, brushing, and cleaning the former underside—but not the exposed side—of his framed fossil, until at last the great fish lay exposed in reverse from its original position. The underside, too, was quickly sealed with shellac and otherwise prepared for exhibition with minor reconstruction of broken or missing bone parts.

George employed a method he had perfected as a trademark or signature for all his wall-mounted specimens: using a tool somewhat like a meat tenderizer with sharp points, he tapped the plaster of the background of the specimen to create a pebbled or grained effect. Finally, his big fish was finished, mounted, ready for sale. It attracted widespread publicity, and he soon sold it to the Denver Museum of Natural History.

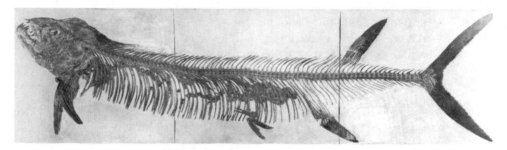

Sternberg sold this Portheus *with fragments of a* Gillicus *inside to the Denver Natural History Museum.*

The next few years were relatively uneventful. One summer he went to New Mexico on a two-month expedition with the Smithsonian, then to Wyoming where he worked with the institute's Dr. Gazin. In the summer of 1949 he joined a field party at Green River, Wyoming, working for the Royal Ontario Museum. In the party were Dr. Loris Russell of the museum and Gus Lindblad, Levi Sternberg's brother-in-law, who had been a member of the Sternberg party in Canada thirty years earlier.

They found a skeleton of *Hyracotherium* (*Eohippus*) or dawn horse only about one foot (30 cm) high and about twenty million years old, George calculated.

George realized the necessity of keeping the paleontologists of the country informed about his discoveries and sent brief reports regularly to the *News Bulletin of the Society of Vertebrate Paleontology*. Such reports often led to the sale of his fossils and also directed the attention of the society's members to the development of the museum at Hays.

It was always easier to get publicity for discoveries of large mammals or great fish, but Sternberg also collected numerous small specimens, such as he described in one issue of the *News Bulletin* where he reported finding a nearly complete limb and foot of the flying reptile *Pteranodon* and several bones of a small bird. George wrote:

> I am glad to report that the fine *Inoceramus* shell which I found in May of '48 has been taken over by our own college museum along with a partial skeleton of the pterodactyl *Pteranodon*. The specimen of *Pteranodon* has a wing expanse of about 22 feet [6.6 m]. We have one complete wing, part of the opposite one and most of the body elements but only one femur of the hind limbs. The skull had been entirely destroyed when found, but recently we secured a fair skull of another individual which we believe will go with this specimen. We are waiting for more parts of other individuals in hope of making a composite mount of this rare reptile.
>
> A third specimen, one of the mosasaur *Platycarpus coryphaeus,* was collected from near Gove City, Kansas, and is now being prepared. It is 18 feet [5.4 m] long from tip to tip and every vertebra seems to be present down to the very last one. A fore and hind paddle are complete and in position, while the other two are present but scattered. It will be necessary to remove the ribs and straighten them out, but it is hoped that a very fine slab mount will emerge from this very complete skeleton.[19]

All the furies of man and nature seemed to crash around the Sternbergs at one time or another. But in each case, they courageously rose above the threatened calamity. George's father, Charles H., had encountered Indians and witnessed the destruction of his fossils by tornado. In Canada father and sons were in peril when their raft was in a flood. More recently, George had suffered through dust, drought, and a prairie fire.

George's final conflict with the forces of nature came on May 22, 1951 when a cloudburst with eleven inches (28 cm) of rain near Hays sent a normally gentle, meandering stream raging through the little city in the predawn darkness, causing death and destruction in a matter of minutes. Six people drowned, and many homes were badly damaged. The college campus was inundated; the building housing the museum was one of the hardest hit. The museum was at ground level with fine exhibits from the natural history and Western collections placed near the floor. Many specimens, records, and other irreplaceable materials were destroyed, and the building was badly damaged.

Sternberg's home sustained damage. Only two blocks down the street in a flooded house, Dr. Wooster's stepson drowned trying to salvage photography equipment stored in the basement. But the Hall of Paleontology at the museum had no major losses, and by fall the campus had been restored.

A few months after the flood, in the fall of 1951, a farmer was making a new road grade in southeastern Gove County when his bulldozer struck fossil bones—the end portions of a radius, an ulna, and a tibia—shearing off the extreme ends of each. The bones extended downward at a thirty degree angle into soft clay, and each had a complete set of articulated foot bones attached. The 42-inch (107-centimeter) long radius and ulna and 32-inch (81-centimeter) long tibia indicated that the animal was no small creature.[20] It was, in fact, a mammoth, but Sternberg, who was called to investigate and evaluate the discovery, said it wasn't a great find, laughingly adding that if it was less than 150 million years old, he could't really get excited. He did, however, help recover the fossil. It was not of great value, but became an object of curiosity for the farmer who kept it.[21]

George Sternberg counted three great highlights in his lifetime. They represented three different periods of his fossil-hunting career. The first was the "mummy dinosaur" (now in the American Museum of Natural History in New York) he found when working with his father and brothers in Wyoming in 1908. The second was the strange little pachycephalosaur *Stegoceras* or "dome-headed" dinosaur (now in the University of Alberta Museum in Edmonton) he found in Canada in 1921. The third was another—the final and most famous— "Fish-Within-a-Fish" removed in 1952, about forty-five miles (72 km) southwest of Hays in Gove County.[22] This would remain in George's "own" museum in Hays, Kansas, near its discovery site.

Myrl Walker loved to tell the story of how the great fossil fish happened to stay in Kansas.[23] In the spring of 1952 Dr. Bobb Schaeffer and Walter Sorenson from the American Museum were traveling to the Western states and stopped off in Hays to visit George and the college museum. Dr. Schaeffer often stopped in Hays and occasionally went with Sternberg to the chalk beds. This time he was introducing Sorenson to the fossil beds of the area, as well as acquainting him with Sternberg.

The three men headed for the badlands, finding only minor items, when suddenly Sorenson saw large flattened bone-like material along a small erosion channel in the rock. He summoned Sternberg, who quickly identified it as the exposed lobe of the tail fin of a *Portheus*. The trio uncovered a few feet of the specimen and realized it would take considerable work to remove the overburden. They also deduced, by following the vertebral column as far as it had been uncovered, that the bones were somewhat disturbed. Possibly the specimen was not really very good. Probably some parts were missing or scattered. At any rate, they didn't have time that day to continue the digging; they made sure the fossil was completely covered by rocks, set up a small marker a short distance from the spot to facilitate later recognition, and returned to Hays.

There are unwritten laws and common courtesies in all businesses. Among fossil hunters, the law was that any discovery of any significance becomes the property of the discoverer. Sorenson found the specimen. Presumably it would be claimed by him for the American Museum of Natural History. However, after considerable discussion and reasoning, added to the fact his museum already had a good specimen of a *Portheus*, Dr. Schaeffer decided his museum would not care to spend further money or time to excavate this particular fossil. He informed Sternberg of the decision and suggested George might want to investigate further and perhaps collect it for himself.

George bided his time, reasonably sure the location would not be disturbed. He had no legal claim to any fossil site. He always had a working arrangement with landowners who granted him permission to hunt whenever he wished and gave him full possession of any fossil he might find. Unlike mineral rights, "fossil rights" were loosely defined and typically handled with a handshake. Like his father and brothers, George never paid landowners for access to their land or for fossils found there, but he never abused his privileges. He respected gates and fences, livestock, and property in general. He always reported any discovery to the landowner immediately and credited the owner when reporting the discovery to the press or to a museum.

A few days after Sorenson and Dr. Schaeffer left, George returned to the fossil site alone and began chipping away at the chalk rock with his prospector's pick. As he began to see the bones, he changed to small awls, and then to soft brushes. The bones were dark and easily discernible. To his surprise, the vertebral column with ribs attached continued deep into the bank of rock. The little irregularity noted earlier was of no consequence. He kept uncovering more of the skeleton, marveling all the time at the completeness—the total skull and lower jaws in place, with the large shiny teeth that characterized the species. Although particles of rock still covered parts of the skeleton, he could nevertheless tell it was all there!

To his amazement, he realized he had uncovered about fourteen feet (4.2 m) of fish; and, while this was not the largest he had ever found—some had been as long as seventeen feet (5.1 m)—it was perfect, complete!

His thoughts turned again to Dr. Schaeffer, Sorenson, and the American Museum. No doubt, if they had known the magnitude of their discovery they might not have given it up so quickly. Sternberg was a man of integrity: he covered his fish with canvas, camouflaged it with rocks, and drove back to Hays to telephone Dr. Schaeffer and again offer him the chance to claim it.

Dr. Schaeffer considered the travel distance and shipping costs, plus his other commitments for the summer, and finally told Sternberg to go ahead. The fossil was his. Good luck! He was free to collect it, prepare it, and eventually offer it for sale to a museum that needed such a specimen.

With a free conscience and light heart, Sternberg went back to the site to uncover his fish. Starting at the tail, he worked slowly up the long spine, carefully cleaning each vertebra and bone. If, as sometimes happens, a bone broke or chipped, he would immediately glue it back in place with a special acetone solvent glue that quickly filled and hardened the bone in place. He had gone about three feet (91 cm) and was moving toward the pelvic fin when he saw beneath the ribs of his fossil what appeared to be the small skull of another fish. He had seen remnants of small fish within other specimens, and remembered his big fish of several years earlier, lying in perfect position, with part of a smaller fish within. But this was much superior, a perfect skeleton of a fish, approximately six feet (1.8 m) long, beautifully visible within the carcass of the larger fish. The tail of the small *Gillicus* was almost up to the throat of the *Portheus*. The vertebrae column was straight and absolutely in place. The head was between the ribs of the right and left sides of the big fish. The evidence was there: the fourteen-foot (5.2-centimeter) fish had swallowed the smaller fish, and together they had died and settled to the bottom of the sea.

Again, Sternberg felt it was his duty to call Dr. Schaeffer. Again, he covered the fossil, headed back to Hays and phoned New York. Again, after due consideration, he was assured that the fossil was his, and that he could go ahead and collect it!

He pitched a tent near the quarry when he first began working on the fossil, feeling the necessity to stay close by. Word spread that he had found a great fish, and curiosity seekers as well as reporters and genuinely interested parties were finding their way to the site. George became apprehensive, fearing someone would inadvertently step on the fish and damage the fragile bones. Tension mounted: he was anxious to get the fossil safely removed. Besides, summer had moved in, and summers in the rock beds of western Kansas often bring blistering heat. George was now nearly sixty-seven years old, and his endurance was not quite as good as it once had been.

By the end of June the fossil was clean and ready for the wooden frames and the plaster. With the Fourth of July holiday coming up, George took advantage of the long weekend and planned a field party for his faculty friends. There was much to be accomplished, and timing was crucial. Proper consistency of the plaster was essential, and temperatures altered its consistency and the time required for hardening it.

To ensure the correct degree of solidity, the plaster was hand mixed in several large tubs, then poured quickly to prevent premature drying or variation in the different batches of mix. A prepared mixture of dextrine, a corn sugar derivative, was added to the water to aid the setting texture and hardness. Drying too quickly could be as disastrous as drying too slowly. Several large cakes of ice were on hand to keep the water cool. Finally George poured the plaster over the bones, filling the two big wooden frames to a depth of about four inches (10 cm). Then he spread a large canvas tarp over the frames and sprinkled it with water to keep the slabs cool as the plaster slowly hardened. Then came the waiting period of a few hours before the covers were nailed on the frames. Next came the trenching and undercutting so the slabs could be lifted and turned bottoms up.

The removal of the huge fossil made for a Fourth of July celebration the little group of weekend fossil hunters would never forget. When the heavy slabs were upended and ready to load for the trip to Hays, Sternberg lined the truck bed with old automobile tires to cushion the

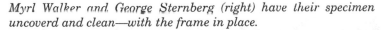

Myrl Walker and George Sternberg (right) have their specimen uncoverd and clean—with the frame in place.

slabs and prevent jarring and possible fracture. The whole operation was executed without any unfortunate incident, and the slabs were carried into the museum laboratory and placed on long tables to await the final preparation.

It had been a full month since Sternberg began chipping out the fossil. The whole operation was on film, thanks to Dr. Wooster, and the story appeared in the press nationwide at the time of removal of the fossil, and again in more detail, some months later when it was ready for exhibit.

For all the helpers, the job was done when the slabs were delivered to Sternberg's workroom. But for George there remained the painstaking work of preparation. In the lab he began chipping away the new topside, thus revealing the other side of the fish. Speed became important because he knew the necessity of working while the slab was still moist. Even so, occasional "watering down" was necessary. It was a hot summer, and there was no air conditioning in the museum.

When at last George wiped away the final particles of loose rock with a damp sponge, there lay his prize in full glory, ready for minor bits of repair and finishing touches. The slabs, still in their frames, were allowed to dry thoroughly. Then George took his special tool and chipped at the background plaster to create his trademark textured

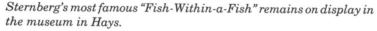

Sternberg's most famous "Fish-Within-a-Fish" remains on display in the museum in Hays.

effect. He applied a final color to all exposed plaster. The bones, of course, had long since received several coats of shellac to preserve and harden them. The final step involved mounting the two slabs, fitting them together at the center as closely as possible, and then filling in the crack with plaster to create a solid one-slab display. An oak frame matching the wood trim of the museum rooms encased the entire slab, and the "Fish-Within-a-Fish" was ready for display.

It was autumn by the time Sternberg had his fish within a fish on the museum wall. He needed to sell it, but he wanted to exhibit it first. The new president of the college, Dr. M. C. Cunningham, had no funds available to make the purchase. But George had an idea. There was another big *Portheus* on the wall, the one former President Lewis had bought for $1,200 in 1927 when Sternberg first came to Hays. This museum really didn't need two large exhibits of a *Portheus*, and George could not bear to part with his new fish.

He proposed to trade the fish within a fish for the old *Portheus*, which he would then sell to some other museum needing such a fossil. Obviously, the extra money he could have claimed for selling his new prize was not as important as keeping it in "his" museum. This way he could recover his expenses and still have the big fish where he could enjoy showing it. President Cunningham was delighted and quickly accepted the offer. The other *Portheus* was soon sold to the State Museum of the University of Nebraska.

He had found other examples of a fish within a fish, with both carcasses recognizable, but this was his crowning achievement and one which was heralded around the world as unique. It has been claimed by paleontologists and textbook publishers that this fossil is the most widely photographed fossil in the world, appearing in countless school texts as well as in other books on paleontology and natural history.

Ever since joining the staff at Hays, Sternberg had followed a pattern of placing his new specimens in the museum, and then, when he'd found a better example of the same kind, selling the lesser selection, so he could always keep the best. This led to continual swapping, resulting in an ever-increasing supply of new and more select specimens for the college. If George discovered he had three of any one kind, he would sell the lesser two and keep the finest. This satisfied his ego, for he was possessive of the museum. He swapped not so much out of love for the college, but out of the desire to please himself and to keep the best in sight, according to his co-workers. But ultimately, the college museum was the benefactor; for it was unofficially "Sternberg's Museum" as long as he lived, and his pride in it was unsurpassed.

In his desk he kept a list of other museums interested in his fossils. He didn't try to make a lot of money on any one item. He figured the number of days he'd spent and a fair daily wage for his time, plus his actual expenses. Then he'd add a little "margin" and that was his price. He had been taught honesty, integrity, and fair play, and he took advantage of no one.[24]

National Geographic, Life, the *London Times*, and other publications picked up the story of Sternberg's most magnificent "Fish-Within-a-Fish." Its picture has been in children's books, and in other books on travel, adventure, and science. Museum directors from all over the world and uncounted thousands of visitors have found their way to Hays to see the amazing fossil. It was indeed a fitting climax to mark the era of Sternberg fossil hunting.

Sternberg at rest!

The Sternberg Memorial Museum

N othing ever seemed as exciting to George Sternberg as securing his most famous "Fish-Within-a-Fish" in 1952. For the next decade he continued to hunt, but his trips became shorter and less productive; and he spent more time developing the museum that became his deepest source of pride.

In the summer of 1961 at age seventy, he was forced to retire because of the mandatory age law. Myrl Walker was immediately named his successor, but Sternberg continued to spend time in the field and the museum building. In 1962 he attended the annual meeting of the Society of Vertebrate Paleontology in Chicago, and that year found and sold a *Pteranodon* specimen to the University of Nebraska State Museum and a large *Portheus* skull to the University of Arizona.

He continued to make entries in his field books, and one of his last entries, dated May 13, 1963, concerned the fish *Saurodon*. He sketched the discovery in its plaster crate as fifty-four inches (137 cm) long and eighteen inches (46 cm) wide. He made this comment: "Reinforced wood pieces with plaster and burlap strips. This was put on in workshop and will be done when finished." George needed no audience for his occasional humorous comments: he wrote for himself.

Shortly thereafter he suffered a slight stroke that left him with diminished mental and physical capacities. He spent his last two years living in a nursing home in Hays and died on October 23, 1969. As had happened with his father, tributes poured in from across the nation upon his death. Dr. David Bardack of the University of Illinois wrote: "George was a contact to people and places of the remote paleontological past. I shall never forget the several delightful times with George running about the chalk beds." Dr. Donald Baird, of the Department of Geology at Princeton University, said, "He was a man who left his mark and will be remembered as long as bone is joined to bone in the museums of the world." Dr. Alexander Wetmore of the Smithsonian wrote: "He was a fine man in every way, devoted to his science and one who made definite contribution to it."

Walker was instrumental in reorganizing the museum and having it officially designated in 1970 as The Sternberg Memorial Museum to honor all members of the family of fossil hunters.

Portraits of four Sternbergs hang in the museum—the Reverend Levi, Dr. George M., Charles H., and George F. A small showcase holds special items representative of the work of the Sternbergs: a doctor's bag and medical instruments of the Civil War period, such as Dr. George M. would have carried; personal items and tools used by Charles H. and a first-edition copy of his autobiography, *Life of a Fossil Hunter*; George F.'s passport for his trip to Patagonia and the little tool he designed to create his trademark, the pitted texture on

All buildings on the Fort Hays State University campus are constructed of western Kansas stone imbedded with fossil shells. In this close-up of the area at the right of the entrance to the Sternberg Memorial Museum, fossil formations are visible in the upper center and to the far right. —Charlie Riedel

Sternberg Museum has occupied the first floor of McCartney Hall on the campus of Fort Hays State University since George Sternberg assumed the post of curator in 1927. —Fort Hays State University

the plaster back base of the specimens he prepared for exhibit. The largest item in the case is the hunk of Niobrara chalk George inscribed when, at age nine, he found his first fossil.

The Sternberg Museum has five halls—Paleontology; Natural History; Archaeology and Ethnology; History; and Pioneers. Another large area is designated for the geology collection, with displays of fluorescent minerals, meteorites, and the three major classes of rocks: igneous, sedimentary, and metamorphic.

The museum houses approximately three million specimens and artifacts, of which about one percent are on display at any one time. Between two and three thousand of the fossils on exhibit were collected by Sternbergs.

Visitors to the Hall of Paleontology usually focus first on the wall to their left as they enter: here hang some of George F.'s best known specimens—a big slab of crinoids, a huge turtle, and the "Fish-Within-a-Fish." At the far end of the hall the visitor sees the giant mosasaur *Tylosaurus proriger*, discoverd by Charles H. and prepared by George. A *Pteranodon* seems to hover above other exhibits as the eye traverses the room filled with showcases for ancient horses, mammoths, and other prehistoric fossils. A fossil shark reveals 200 teeth and 210 vertebrae with a somewhat "mummified" body eighteen feet (5.4 m) long.

One case features items from Canada: the cast of the dome-headed dinosaur *Stegoceras*, found by George in the Belly River beds. The cast was prepared by Levi and presented to the museum by Charles M., who also presented a footprint cast of a dinosaur he found in the Peace River Canyon of British Columbia. Trilobites from Hull, Ontario, and other items remind visitors of the years the Sternbergs spent in Canada.

At this time, there are no large open mounts nor complete dinosaur skeletons in the museum. The main focus has been on the discoveries from the Niobrara chalk beds, and dinosaur remains are not common in Kansas.

Visitors come to the museum from all over the world; more than twenty thousand register each year, typically representing more than twenty countries, as well as all the contiguous forty-eight states. To aid students of all ages, the directors seek to explain the geology and also the processes involved in discovering and preparing fossils. Through these explanations and displays, they bring the story of the past to life; with guided tours, visitors can see and study these clues to life millions of years ago.

Early in 1991 Fort Hays State University announced the acceptance of a gift from the Chrysler First Business Credit Corporation— a Hays building complex containing 88,747 square feet (8,245 sq m) and valued at $4.5 million. The complex is earmarked to become the new home of the Sternberg Museum. The larger premises will enable the university to expand the museum so that its holdings can be displayed more effectively and so that more traveling exhibits can be brought to Hays. The space holds vast educational and research possibilities, beginning with the move of the current museum from its on-campus home to its new location just off Interstate 70, approximately two and a half miles (4 km) from the university.

Plans call for transforming the museum into a world-class educational facility and tourist attraction, with hands-on physical science exhibits explaining light, sound, heat, magnetism, electricity, and other physical and technological phenomena and innovations. Exhibits will focus on biology and natural history, with displays representing the components of living organisms and their function, diversity, and evolution, and others pertaining to weather and climate. One area will feature a planetarium. A Discovery Room will enable children to make hands-on investigations.

Artist's sketch of the entrance to the domed structure and attached building that will be the new home of the Sternberg Museum at Fort Hays State University. —D. Moore, Fort Hays State University

Planners hope that one floor of the building will be able to have as its centerpiece several mechanical, lifelike, mobile and vocal dinosaurs, pteranodons, plesiosaurs, and mosasaurs, with an exhibit of large mammals from the glacial ages nearby.

Exhibit design and production facilities, a gift shop, library, conference room, and an auditorium for public education programs, science demonstration and multimedia presentations, should make the Sternberg Museum a major learning center in the heart of America—very near the source of many of the finest fossils displayed.

University officials hope that that museum can be in its new home by 1995.

With the death of George and his brothers, the Sternberg era of fossil hunting ended. None of the younger generation of Sternbergs cared to continue in their path. Perhaps the intensity of the male-dominated family passion for fossil hunting was unacceptable to the twentieth century Sternbergs. Or perhaps it was just a natural change with the times.

The Sternberg dynasty, like the dinosaurs themselves, could not last indefinitely. But even though no Sternbergs are actively carrying on the family fossil-hunting tradition, all those who visit "their" museum and all who respect their adventure, determination, and courage to recover life from the past, are heirs to their dinosaur dynasty.

Notes

Chapter One

1 Female students were not accepted at Hartwick at that time.

2 Charles H. Sternberg, *Life of a Fossil Hunter* (New York: Holt and Co., 1909) 2.

3 Ibid.

4 Robert West Howard, *The Dawnseekers* (New York: Harcourt, Brace, Jovanovich, 1975) 164-68.

5 Ibid., 5-14.

6 Ibid., 23-25.

7 Ibid., 28-31.

8 Craig Miner, *West of Wichita, Settling the High Plains of Western Kansas 1856-1859* (University Press of Kansas, 1986) 35.

9 Ellsworth County, Kansas, Register of Deeds Office archives.

10 Grace Muilenberg and Ada Swinford, *Land of the Post Rock; Its Origins, History, and People* (University Press of Kansas: 1975) 12.

11 Miner, *West of Wichita*, 35.

12 Ibid., 58.

13 Ibid., 38.

14 Ibid., 51

Chapter Two

1 Sternberg, *Life of a Fossil Hunter*, 10-11.

2 Robert Utley, ed., *Life in Custer's Cavalry, Life and Letters of Albert and Jennie Barnitz* (New Haven: Yale Press, 1977) 59

3 John M. Gibson, *Soldier in White; the Life of Gen. George Miller Sternberg* (Durham, N. C.: Duke University Press, 1958) 36.

4 Dr. Leo Lesquereux identified these leaves as sassafras, but in recent years the Smithsonian has classified them as "Cretaceous relatives of sycamores," according to the Smithsonian's Scott Wing, writing to the author in 1990.

5 Sternberg, *Life of a Fossil Hunter*, 14.

6 Archives of the Registrar's Office, Kansas State University, Manhattan, Kansas.

7 Ibid.

8 Sternberg, *Life of a Fossil Hunter*, 20.

9 Gibson, *Soldier in White*, 37.

10 Ibid., 36.

11 Sternberg, *Life of a Fossil Hunter*, 17.

12 Ibid, 19.

13 Ibid.

14 Ibid., 20-24

15 These discoveries were recognized by Dr. Arthur Hollick in a paper, "A Fossil Petal and a Fruit from the Cretaceous (Dakota Group) of Kansas," in *Contributions from the New York Botanical Garden*, no. 31.

16 Sternberg, *Life of a Fossil Hunter*, 5.

17 *The Ellsworth Reporter*, May 28, 1872.

18 The little church, now a chapel and part of the Ellis County Kansas Historical Society's holdings, is on the National Register of Historical Places. Evidence of fossils is still visible. Five blocks west of the church is the campus of Fort Hays State University, proof that Martin Allen was more successful than Sternberg in getting government land turned over to the state. All campus buildings are of native stone and have plainly visible embedded fossils.

Chapter Three

1 Robert Plate, *The Dinosaur Hunters* (New York: David McKay and Co., 1964) 40-47.

2 Howard, *Dawnseekers*, 10-21.

3 Plate, *Dinosaur Hunters*, 27-39.

4 Ibid., 60-62.

5 Ibid., 94.

6 Ibid., 100.

7 Ibid., 55-56.

8 Howard, *Dawnseekers*, 132-40.

9 Sternberg, *Life of a Fossil Hunter*, 32-34.

10 Ibid., 33.

11 Ibid, 36.

12 Ibid, 37.

13 Ibid., 40-41.

14 Ibid, 44.

15 Ibid.

16 Ibid., 47-48.

17 Ibid., 61.

18 Ibid.

Chapter Four

1 Sternberg, *Life of a Fossil Hunter*, 61.

2 Plate, *Dinosaur Hunters*, 170.

3 Sternberg, *Life of a Fossil Hunter*, 64.

4 Ibid., 65.

5 Ibid., 66-67.

6 Plate, *Dinosaur Hunters*, 171-72.

7 Sternberg, *Life of a Fossil Hunter*, 70.

8 Ibid., 76.

9 Plate, *Dinosaur Hunters*, 175.

10 Sternberg, *Life of a Fossil Hunter*, 82-83.

11 Ibid., 87.

12 Howard, *Dawnseekers*, 230-38.

13 Plate, *Dinosaur Hunters*, 177-78.

14 Ibid.

15 Ibid., 179.

16 Ibid.

17 Sternberg, *Life of a Fossil Hunter*, 98.

Chapter Five

1 Sternberg, *Life of a Fossil Hunter*, 99-100.

2 Letter from the files of George F. Sternberg, Forsyth Library Archives, Fort
    Hays State University, Hays, Kansas.

3 Sternberg, *Life of a Fossil Hunter*, 120.

4 Ibid., 123.

5 Ibid., 142.

6 Ibid., 141-42.

7 Elizabeth Noble Shor, *Fossils and Flies* (Norman: University of Oklahoma
    Press, 1971) 87.

8 Sternberg, *Life of a Fossil Hunter*, 151.

9 Ibid., 158.

10 Ibid., 168.

11 Gibson, *Soldier in White*, 87-90.

12 Ibid.

13 Ibid., 95-110.

14 Sternberg, *Life of a Fossil Hunter*, 173-80.

15 Alfred Romer, *Vertebrate Paleontology* (University of Chicago Press, 1965) 281.

16 Sternberg, *Life of a Fossil Hunter*, 188-89.

17 Ibid., 188.

18 Ibid., 191-200.

19 Ibid., 203-4.

Chapter Six

1 Letter from Charles M. Sternberg to Myrl V. Walker, December 20, 1977, Forsyth Library Archives.

2 Sternberg, *Life of a Fossil Hunter*, 126-28.

3 Ibid., 128-29.

4 Ibid., 137-38.

5 Ibid., 129-33.

6 Ibid., 205.

7 Ibid., 205-7.

8 Ibid., 207.

9 Ibid., 213.

10 Ibid., 217.

11 Ibid., 225.

12 Ibid., 227.

Chapter Seven

1 Sternberg, *Life of a Fossil Hunter*, 129-34.

2 Shor, *Fossils and Flies*, 109.

3 Sternberg, *Life of a Fossil Hunter*, 133.

4 Shor, *Fossils and Flies*, 108.

5 Sternberg, *Life of a Fossil Hunter*, 133.

6 In *Fossil Feud* (Hicksville, N.Y.: Exposition Press, 1974), author Elizabeth Shor devotes much space to the Cope-Marsh rivalry and its effect on all fossil hunters and the museums that bought their discoveries.

7 *New York Herald*, January 19, 1890.

8 Shor, *Fossils and Flies*, 78.

9 One of many stories about the life George F. Sternberg, preserved by Myrl V. Walker, his friend and colleague, and related to the author on August 30, 1983.

10 Shor, *Fossils and Flies*, 140.

11 Ibid.

12 George F. Sternberg, "Thrills in Fossil Hunting," *The Aerend, a Kansas Quarterly*, (vol. 1, no. 3, Summer 1930), 139-46.

Chapter Eight

1 Sternberg, *Life of a Fossil Hunter*, 231-32.

2 Ibid., 233.

3 Ibid., 234-36.

4 Ibid., 241.

5 Ibid., 243.

6 The list of these works fills 145 pages of fine print in Henry Fairfield Osborn's biography, *Cope: Master Naturalist*.

7 Ibid., 109.

8 Ibid., 112.

9 Ibid., 54-57.

10 Ibid.

11 Ibid., 58.

12 Ibid., 244.

13 Ibid., 252.

14 Ibid., 253.

15 Ibid., 255.

16. Dr. Broili was paying Sternberg $200 a month for this expedition, as shown by Sternberg in a letter from Broili, December 23, 1901, reproduced in *Life of a Fossil Hunter*. Broili also paid freight expenses: $3.46.

17 Sternberg, *Life of a Fossil Hunter*, 256-57.

Chapter Nine

1 Sternberg, *Life of a Fossil Hunter*, 114-16.

2 Ibid., 117.

3 Ibid., 117-19.

4 George F. Sternberg, "Thrills of Fossil Hunting," 147.

5 Ibid.

6 Sternberg, *Life of a Fossil Hunter*, 270.

7 Ibid., 278-81.

8 Ibid., 267.

9 Ibid., 266.

Chapter Ten

1 Sternberg, *Life of a Fossil Hunter*, 270-71.

2 Alfred S. Romer, *Vertebrate Paleontology*, 3rd ed. (Chicago: University of Chicago Press, 1966), 37-42.

3 Sternberg, *Life of a Fossil Hunter*, 272.

4 Charles Sternberg carefully avoided quoting prices he received for his fossils or revealing financial arrangements with museums.

5 According to Sternberg, *Life of a Fossil Hunter*, 270-274, later the Sternbergs took out of this bed a wide assortment of teeth of reptiles and fishes; scales of ganoid, or gar, and sturgeon; bones of small dinosaurs and crocodiles; and beautiful sculptured shells of turtles.

6 George F. Sternberg, "Thrills of Fossil Hunting," 135-53.

7 Ibid.

8 Ibid.

9 Sternberg, *Life of a Fossil Hunter*, 275.

10 Douglas Preston, "Sternberg and the Dinosaur Mummy," *Natural History* (1982) 91:88.

11 Henry F. Osborn, "Introduction" to 1909 edition of Sternberg's *Life of a Fossil Hunter*.

12 By 1931 interest in the book was still so strong that Charles revised it for a second printing. This edition has been widely quoted and used as a reference by numerous writers about paleontology and the notorious Marsh-Cope feud.

13 Charles H. Sternberg, *Hunting Dinosaurs in the Badlands of the Red Deer River*, rev. ed. (San Diego: Jensen Press, 1932), 4.

14 Ibid., 4-5.

15 Ibid., 5-7.

16 Ibid., 7.

17 Charles H. Sternberg, "Fossil Monsters I Have Hunted," *Popular Science Monthly* (December 1929), 56.

18 Sternberg, *Hunting Dinosaurs*, 10.

19 Ibid., 10-11.

Chapter Eleven

1 Charles H. Sternberg, "Introduction," *A Story of the Past, or The Romance of Science* (Boston: Sherman, French and Co., 1911).

2 David A. E. Spalding, "Introduction" to Sternberg's *Hunting Dinosaurs in the Badlands of the Red Deer River*, 3rd ed. (Edmonton, Canada: NeWest Press, 1985), xxvii-xxix.

3 Charles H. Sternberg, *Hunting Dinosaurs*, 9-10.

4 Ibid., 13-14.

5 Incident retold by George F. Sternberg to Myrl V. Walker, related to author September 2, 1952.

6 Sternberg, *Hunting Dinosaurs*, 11-12, quoting March 1, 1913 article in *London Illustrated News*.

7 Ibid.

8 Ibid., 14.

9 Casts of this specimen grace many state museums in Europe as gifts of Andrew Carnegie, who once cabled Hatcher asking the cost of making a plaster restoration of the specimen. When Hatcher quoted $10,000, Carnegie immediately placed the order so he could present it to the British Museum.

10 Gibson, *Soldier in White*, 172-80.

11 Personal letter to Myrl V. Walker from Dr. John C. Stoltz, June 17, 1919, Sternberg papers (Forsyth Library, Fort Hays State University, Hays, Kansas).

12 Gibson, *Soldier in White*, 36.

13 Copy of last will and testament of George Miller Sternberg, Sternberg papers (Forsyth Library, Fort Hays State University, Hays, Kansas).

14 Sternberg, *Hunting Dinosaurs*, 20-22.

15 Ibid., 25.

16 Ibid., 26.

17 Ibid., 27-28.

18 Ibid., 28-32.

19 Ibid.

Chapter Twelve

1 Edwin H. Colbert, *Men and Dinosaurs* (New York: E. P. Dutton & Co., 1968) 186.

2 Ibid., 187.

3 Brown's collections from the Red Deer River from 1910 to 1915, together with those made previously from the Hell Creek beds of Montana, made it possible for the American Museum of Natural History to have the finest display of Cretaceous dinosaurs in the world.

4 Colbert, *Men and Dinosaurs*, 186.

5 Ibid., 187.

6 Ibid., 191.

7 Sternberg, *Hunting Dinosaurs*, 40-42.

8 Ibid.

9 Ibid., 44-46.

10 Jack McGee, like many Sternberg assistants, just appears and disappears. Quite possibly Jack was a son of Dan McGee.

11 Sternberg, *Hunting Dinosaurs*, 49.

12 Ibid., 50-53.

13 Ibid., 51.

14 Ibid., 56.

15 Ibid., 61-62.

16 Ibid., 82.

17 Colbert, *Men and Dinosaurs*, 195.

18 George Sternberg photo albums, Forsyth Library Archives, Hays, Kansas.

19 Sternberg, *Hunting Dinosaurs*, 85-86.

20 George Sternberg photo albums.

21 Ibid.

22 Charles H. Sternberg, copy of letter to Lambe, October 13, 1913, Forsyth Library Archives.

23 Ibid.

24 Lawrence Lambe's *Report to Geological Survey*, 1913, 293.

25 Sternberg, *Hunting Dinosaurs*, 107-8.

26 Howard, *Dawnseekers*, 267.

27 Loris S. Russell, "Dinosaur Hunting in Western Canada," *Life Science*, (Undated), Royal Ontario Museum, Contribution 70, 19.

28 The citizenship status of the Sternbergs at this time is uncertain. There is some question as to why none of the brothers returned to the United States until after the end of World War I.

29 Russell, "Dinosaur Hunting," 20.

30 The dream takes up more than fifty pages of Charles Sternberg's *Hunting Dinosaurs in the Badlands of the Red Deer River*.

Chapter Thirteen

1 Loris S. Russell, "Tribute to Charles Mortram Sternberg," *Canadian Field Naturalist* 1982, 483.

2 Sternberg, *Hunting Dinosaurs*, 189-90.

3 Ibid., 190-91.

4 Sternberg, "Fossil Monsters I Have Hunted," 55ff.

5 Other expeditions were sent out between 1920 and 1962, and collectors found specimens in the valleys of the Bow, Little Bow, and Milk Rivers, along the western Cypress Hills in Southern Alberta and in the valleys of the South Saskatchewan River. However, none of these collections were large or particularly significant. Only a few notable specimens were discovered after this date, but there was a great need for scientific study of what had been found and preserved, and for developing the Canadian museums to serve the country better.

6 Russell, "Dinosaur Hunting," 21.

7 Ibid.

8 Sternberg, "Thrills of Fossil Hunting," 149-50.

9 Russell, "Dinosaur Hunting," 22.

10 Spalding, "Introduction" to "Hunting Dinosaurs," xxxii.

11 Russell, "Dinosaur Hunting," 27-28.

12 Ibid., 28.

13 Ibid., 30.

14 Ibid.

15 Loris S. Russell and Gord Edmund, Obituary, "Levi Sternberg," *News Bulletin, Society of Vertebrate Paleontologists*, no. 110 (June 1917), Texas Memorial Museum, Austin, Texas, 43.

16 Ibid.

Chapter Fourteen

1 Natural History Division, *Reference List* no. 155, Provincial Museum of Canada, Edmonton.

2 Russell, "Charles Mortram Sternberg, 1885-1981," *Canadian Field Naturalist* 96 (1982), 483-89.

3 Russell, "Charles Mortram Sternberg," *News Bulletin, Society of Vertebrate Paleontology* 124 (February 1982), 70.

4 Ibid., 70-72.

5 Russell, "Dinosaur Hunting in Western Canada," 24-25.

6 Colbert, *Men and Dinosaurs*, 141-42.

7 Russell, "Dinosaur Hunting" 23.

8 Ibid.

9 Ibid., 25.

10 Ibid.

11 Map 969 A., Geological Survey of Canada.

12 Russell, "Dinosaur Hunting," 26.

13 Russell, "Charles Mortram Sternberg, 1885-1981," *Canadian Field Naturalist*, 485.

14. Ibid.

15 Russell, "Charles Mortram Sternberg," *Society of Vertebrate Paleontology*, 72.

16 Letter from Charles M. Sternberg to Myrl V. Walker, Archives, Forsyth Library.

17 Russell, "Charles Mortram Sternberg," *Proceedings of the Royal Society of Canada* series 4, vol. 20 (1982), 134-35.

Chapter Fifteen

1 Sternberg, "Fossil Monsters I Have Hunted," 140.

2 Ibid.

3 Colbert, *Men and Dinosaurs*, 141-42.

4 Sternberg, *Hunting Dinosaurs*, 2nd ed., 193-98.

5 Ibid., 201-2.

6 Ibid.

7 Ibid.

8 Ibid., 212.

9 Ibid., 225-27.

10 Ibid., 221.

11 Ibid., 228-32.

12 Ibid., 242-44.

13 Ibid., 238-41.

14 Ibid.

15 Preston, *Dinosaurs in the Attic*, 74.

16 Charles W. Gilmore, "Obituary, Charles H. Sternberg," *News Bulletin, Society of Vertebrate Paleontologists* no. 4 (February 15, 1944), 12-15.

17 David A. E. Spalding, Introduction to "The Later Work of Charles H. Sternberg," *Hunting Dinosaurs in the Badlands of the Red Deer River* (Edmonton, Alberta: NeWest Publishing) xxxviii-xxxiv.

Chapter Sixteen

1 Patagonia means "the land of the people with long feet." It has only one million residents scattered over 450,000 square miles. Fertile land in the northern area lies flat, permitting farming. But in the southern desert area between the Andes Mountains and the Atlantic Ocean, the primary industry is sheep ranching. The country also produces some iron, coal, and minerals. Oil wells dot the area of Comodoro in the north and the territory of Tierra del Fuego at the lower tip of the country.

2 Larry G. Marshall, "Adventures in Patagonia," *Field Museum of Natural History Bulletin* vol. 49, no. 3 (March 1978), 4-11

3 Colbert, 1968, 252ff.

4 Ibid.

5 Elmer S. Riggs, "Fossil Hunting in Patagonia," *Natural History* 26:5 (1926), 536-44.

6 Riggs, "Work Accomplished by the Field Museum Paleontological Expeditions to South America," *Science* 47:1745 (June 8, 1928), 585-87.

7 Marshall, "Adventures in Patagonia," 4-11.

8 Ibid.

9 Ibid.

10 Ibid.

11 Riggs, "Fossil Hunting in Patagonia," 538.

12 Marshall, "Adventures in Patagonia," 4-11.

13 Ibid.

14 Unless otherwise documented, all material in remaining portion of this chapter was taken from an incomplete, handwritten diary, dated 1923, kept by George F. Sternberg in Patagonia. The diary is now in Forsyth Library Archives in Hays, Kansas.

15 The *Astrapotherium* was similar to a tapir in appearance, but shy, gentle, and nocturnal in habits.

16 *Glyptodons* were related to armadillos, with inch-thick, dome-like shells that had hexagonal plates several inches in diameter. The head of the *Glyptodon* sported a helmet, and its legs were short, sheathed in bony plates. The tail was huge and club-like with overlapping bony rings. Since it had no front teeth, it grazed by wrapping its tongue around plants and bringing them into its mouth.

17 Ancient sloths had short, stocky hind legs with three claws and forefeet with huge claws for seizing and pulling down leaves, twigs, and fruit. Since sloths had no front teeth, they depended on long tongues and strong back teeth.

18 Sternberg, "Thrills of a Fossil Hunter," 151-52.

19 Riggs, "Work Accomplished by the Field Museum Paleontological Expeditions to South America," 585-87.

20 Riggs, "Fossil Hunting in Patagonia," 541-43.

21 Ibid.

22 Ibid.

23 This forest has since been designated as a national monument by the Argentine government and is recognized as one of the world's two greatest petrified forests, the other being the Petrified Forest in Arizona.

Chapter Seventeen

1 Charles W. Gilmore to George F. Sternberg, March 31, 1925.

2 Charles H. Sternberg "To Whom It May Concern," February 2, 1925.

3 George F. Sternberg, *Fieldbook*, June 1925.

4 Ibid.

5 Ibid.

6 George F. Sternberg, "Where Birds as Big as Aeroplanes Once Flew," *Union Pacific Magazine* (November 1926) 12-13.

7 Ibid.

8 Ibid.

9 F.W. Irwin and Oakley Board of Education, letter to George F. Sternberg, August 11, 1926.

Chapter Eighteen

1 Myrl V. Walker interview, September 2, 1983.

2 George F. Sternberg, *Fieldbook*, Summer 1927.

3 Ibid.

4 Rudolph Barta interview, June 1986.

5 Walker interview.

6 *Kansas City Star*, April 27, 1930.

7 Walker interview.

8 Ibid.

9 Ibid.

10 Myrl V. Walker interview, August 30, 1983.

11 *Kansas City Star*, July 12, 1925.

12 William (Bill) Eastman interview with author July 2, 1983.

13 Ibid.

14 Ibid.

15 Ibid.

16 Society of Vertebrate Paleontology, *News Bulletin* (February 1950).

17 Bill Eastman remained at the college for ten years and in 1955 left on a sabbatical leave that he spent at the University of Hawaii. During his stint at Fort Hays State, however, he had painted models and scenes depicting a number of major prehistoric animals featured in the college museum displays. After Hawaii, he did research in Africa and Australia, continued his career as a museum artist, and returned to Hays in the early 1980s to paint dioramas for the museum.

18 Society of Vertebrate Paleontology, *News Bulletin* (February 1952).

19 Ibid.

20 Ibid.

21 Walker, interview with author, September 2, 1983.

22 Ibid.

23 Ibid.

24 Ibid.

Appendix

MUSEUMS AND UNIVERSITIES IN
RECEIPT OF STERNBERG FOSSILS
(as listed by Charles H. and George F. on their stationery)

Arizona	Arizona State Museum, University of Arizona, Tucson
Arkansas	The University Museum, University of Arkansas, Fayetteville
California	California Institute of Technology, Pasadena
	Museum of Paleontology, University of California, Berkeley
	Natural History Museum, Balboa Park, San Diego
	Natural History Museum of Los Angeles County, Exposition Park, Los Angeles
Connecticut	Peabody Museum of Natural History, Yale University, New Haven
Illinois	Field Museum of Natural History, Chicago
	Illinois State Museum, Springfield
	Museum of Natural History, University of Illinois, Urbana
Indiana	Joseph Moore Museum, Earlham College, Richmond
Iowa	University of Iowa, Iowa City
Kansas	Pittsburg State University Natural History Museum, Pittsburg
	Sternberg Memorial Museum, Fort Hays State University, Hays
	Systematics Museums University of Kansas, Lawrence
Massachusetts	Agassiz Museum of Comparative Zoology, Harvard University, Cambridge
	Massachusetts Institute of Technology, Cambridge
	The Pratt Museum of Natural History, Amherst University, Amherst

Michigan	Gustavus Adolphus College, St. Peter
	The Museum, Michigan State University, East Lansing
	University of Michigan Exhibit Museum, Ann Arbor
Minnesota	University of Minnesota, Minneapolis
Missouri	Washington University Museum of Paleontology and Geology, St. Louis
Nebraska	House of Yesterday Museum, Hastings
New Jersey	Princeton University Museum of Natural History, Princeton
	Upsala College, East Orange
New York	American Museum of Natural History, New York
	Cornell University, Ithaca
Ohio	Cleveland Museum of Natural History, Cleveland
	Ohio State University, Columbus
	Wittenburg College, Springfield
Pennsylvania	Carnegie Museum of Natural History, Carnegie Institute, Pittsburgh
South Dakota	Augustana College, Sioux Falls
	Museum of Geology, School of Mines and Technology, Rapid City
Texas	University of Texas, Department of Geology, Austin
Washington, D.C.	National Museum of Natural History, Smithsonian Institution
Wyoming	Geological Museum, University of Wyoming, Laramie
Canada	Canadian Museum of Nature, Ottawa
	Royal Ontario Museum, Toronto
	Tyrrell Museum of Paleaontology, Drumheller
	University of Alberta, Edmonton
	University of British Columbia, Vancouver
Denmark	University of Copenhagen Zoological Museum
England	British Museum of Natural History, London
France	Muséum National d'Histoire Naturelle, Paris
Germany	National Museum, Berlin
	Royal Museum, Munich
	Senckenberg Museum, Frankfurt-on-the-Main
	University of Tubingen

Plus many others the Sternbergs listed as being of "lesser importance."

References

I listed here only the major writings that have been useful in compiling this record of the lives of the Sternbergs. This list indicates the range of resources available for those who wish to delve more extensively into the fossil-hunting expeditions of a period now history. The Sternbergs were frequent contributors to scientific and professional journals and were also subjects of a myriad of newspaper and magazine articles. Only a very few of these are listed, as examples of the national attention accorded their activities.

SELECTED BIBLIOGRAPHY

American Council of Learned Societies, ed. *Dictionary of American Biography*. New York: Charles Scribner's Sons, 1935.

Buchanan, Rex, ed. *Kansas Geology*. Lawrence: University Press of Kansas, 1984.

Colbert, Edwin. *A Fossil Hunter's Notebook*. New York: Elsevier-Dutton Publishing Co., 1980.

_____. *Men and Dinosaurs*. New York: E. P. Dutton, 1968.

Colbert, Edwin, and William D. Burns. *Digging for Dinosaurs*. New York: American Museum of Natural History, 1960.

Desmond, Adrian J. *Hot Blooded Dinosaurs*. New York: Dial Press, 1976.

Engleman, Rose C., and Robert J. T. Joy. *Two Hundred Years of Military Medicine*. Ft. Detrick, MD: U. S. Army Medical Department.

Gibson, John M. *Soldier in White: the Life of Gen. George Miller Sternberg*. Durham: Duke University Press, 1956.

Gilmore, Charles W. "Charles H. Sternberg." *News Bulletin of Society of Vertebrate Paleontologists* 4, (15 February 1944), 12-15.

Howard, Robert West. *The Dawnseekers*. New York: Harcourt, Brace, Javanovich, 1975.

Lambe, L. M. "On a New Genus and Species of Carnivorous Dinosaur from the Belly River Formation of Alberta, with a description of *Stephanosaurus marginatuo* from the Same Horizon." Ottawa: *Nature* 28 (1914) 17-20.

Marshall, Larry G. "Bolivian Adventure in Search of the Bones of Giants." Chicago: *Field Museum of Natural History Bulletin* 49: no. 5 (May 1978), 16-23.

_____. "Adventures in Patagonia." Chicago: *Field Museum of Natural History Bulletin* 19: no. 3, (March 1978), 4—11.

McKean, Kevin. "Footprints in Humanity's Past." *Modern Maturity* (February-March 1985), 76-78.

Miner, Craig. *West of Wichita*. Lawrence: University Press of Kansas, 1986.

Monthly, The. "Reminiscences." 8: no. 22, Otsego, New York· Hartwick Seminary (1909).

Nutt, R. Gordon, "Alberta Badland Trails." *Lapidary Journal* 28: no. 5 (August 1974).

Osborn, H. F. "A Dinosaur Mummy." *American Museum Journal* no. 2 (1911).

Plate, Robert. *The Dinosaur Hunters*. New York: David McKay & Company, 1964.

Preston, Douglas. *Dinosaurs in the Attic*. New York: St. Martin's Press, 1986.

_____. "Sternberg and the Dinosaur Mummy." *Natural History* 91, (January 1982), 88.

Riggs, Elmer S. "Fossil Hunting in Patagonia." *Natural History* 91, (January 1982), 536-44.

_____. "Work Accomplished by the Field Museum Paleontological Expedition to South America." *Science* 47, 1745 (June 8, 1928), 585-87.

Russell, Dale. "A Vanished World; The Dinosaurs of Western Canada." *Natural History Series* 4. National Museums of Canada (1977).

Russell, Loris S. "Charles Mortram Sternberg." *News Bulletin, Society of Vertebrate Paleontology* 124, (February 1982).

_____. "Charles Mortram Sternberg." *Proceedings of the Royal Society of Canada* Series 4, vol. 20 (1982), 133-35.

_____. "Charles Mortram Sternberg 1885-1981. *The Canadian Field Naturalist* 96, (1982).

_____. "Dinosaur Hunting in Western Canada 1917-1965." *Life Sciences*, Contribution 70. Royal Ontario Museum. (undated), 14-20.

Russell, Loris S. and Gord Edmund. "Levi Sternberg." *News Bulletin, Society of Vertebrate Paleontology* 110 (June 1977).

Shor, Elizabeth Noble. *Fossils and Flies*. Norman: University of Oklahoma Press, 1971.

_____. *The Fossil Feud*. Hicksville, NY: Exposition Press, 1974.

Spalding, David A. E. Introduction to 3d ed. of *Hunting Dinosaurs in the Badlands of the Red Deer River of Alberta, Canada*. Edmonton: NeWest Press, 1985.

Sternberg, Charles H. *Life of a Fossil Hunter*. New York: Holt & Co. 1909.

_____. *A Story of the Past, or The Romance of Science*. Boston: Sherman French and Co., 1911.

_____. *Hunting Dinosaurs in the Badlands of the Red Deer River, Alberta, Canada*. Published by the author, 1917. Reprint, San Diego: Jensen, 1931. 3d ed., Edmonton: NeWest Press, 1985.

_____. "Five Years' Explorations in the Fossil Beds of Alberta." *Kansas Academy of Science Transactions* no. 28-29, 205-211.

_____. "Fossil Monsters I Have Hunted." *Popular Science Monthly* (December 1929), 56.

_____. Letter/Report to L. M. Lambe, October 13, 1913.

_____. "Were There Proboscis-bearing Dinosaurs?" Annual Magazine, *Natural History Series* no. 2 (1939), 556-556a.

Sternberg, Charles W. "The Dinosaurs of Alberta." *The Aerend, a Kansas Quarterly* vol. 5: no. 3, Hays, Kansas (1935).

Sternberg, George F. "Thrills in Fossil Hunting. *The Aerend, a Kansas Quarterly* vol. 1, no. 3, Hays, Kansas (1930), 139-153.

_____. "Observation of Articulated Limb Bones of a Recently Discovered Pteranodon in the Niobrara Cretaceous of Kansas. *Transactions of the Kansas Academy of Science*. (1958), 81-85.

_____. "Where Birds as Big as Aeroplanes Once Flew." *Union Pacific Magazine* (November, 1926).

Sternberg, George, with Charles H., Charles M., and Levi Sternberg. "In the Niobrara and Laramie Cretaceous." *Kansas Academy of Science Transactions*: 24, (1910).

Sternberg, George, and Erwin H. Barbour. "*Gnathabelodon thorpei*—A New Mud-grubbing Mastodon." *Nebraska State Museum Bulletin* 42: no. 1 (May 1935), 395-404.

Sternberg, George F., and Louis Hussakof. "A New Teleostean Fish from the Niobrara of Kansas." New York: American Museum of Natural History (1929).

Sternberg, George F., and George Robertson. "A Pliocene Waterhole in Western Kansas." *Science* vol. 95, no. 2456 (January 23, 1942).

_____. "Fossil Mammal Tracks in Graham County, Kansas." *Transactions of the Kansas Academy of Science.* no. 45, 1942, 258-61.

Sternberg, George F., and Myrl Walker. "Report on a Plesiosaur Skeleton from Western Kansas." *Transactions of the Kansas Academy of Science* 60: (1957).

Sternberg, Martha L. *George Miller Sternberg.* Chicago: American Medical Association, (1920).

Utley, Robert, ed. *Life in Custer's Cavalry: Life and Letters of Albert and Jennie Barnitz.* New Haven: Yale University Press, 1977.

Walker, Myrl V. "The Impossible Fossil." *University Forum* 26: (Spring 1982), 1-4.

Wooster, L. D. *A History of Fort Hays Kansas State College 1902-1962.* Hays: Fort Hays State College, 1961.

Index

286

287

Weare, Allan 165
Weld County, Colorado 56
Weston, Thomas Chesmer 139
Wetmore, Alexander 256
Whelan, Pat 71
Wichita Falls, Texas 92
Wichita River 85
Wichman, R. 194
Wieland, G. R. 97
Williamson River 54
Williston, Samuel W. 32, 34, 52,
    77–78, 80–82, 89, 170, 194
Willow Springs 91–92
Wiman, Carl 183
Wolfe, R.G. 198
Wooster, L.D. 222–23, 225, 252
World War I 153
    Germans sink fossil-bearing

ship 156
Wortman, "Jake" 59, 62–63
Wyoming 110–21, 126, 223–25, 239.
    See also individual place names

Xiphactinus. See Portheus

Yale 29, 31, 76, 101, 107, 130–31
Yellow Fever Commission 130

Zeigler, Anna Reid. *See also* Sternberg,
    Anna Reid
    marries George F. Sternberg 231
Zoological Museum, University of
    Copenhagen, Denmark 218

288